River-Friendly Cities

Anna Januchta-Szostak

River-Friendly Cities

An Outline of Historical Changes in Relations between Cities and Rivers and Contemporary Water-Responsible Urbanization Strategies

Bibliographic Information published by the Deutsche Nationalbibliothek
The Deutsche Nationalbibliothek lists this publication in the Deutsche Nationalbibliografie; detailed bibliographic data is available online at http://dnb.d-nb.de.

Library of Congress Cataloging-in-Publication Data
A CIP catalog record for this book has been applied for at the Library of Congress.

Publication of this book was financially supported by the Poznan University of Technology and the Ministry of Science and Higher Education in Poland.

Cover image: © Anna Januchta-Szostak

Printed by CPI books GmbH, Leck

ISBN 978-3-631-80151-2 (Print)
E-ISBN 978-3-631-82651-5 (E-PDF)
E-ISBN 978-3-631-82652-2 (EPUB)
E-ISBN 978-3-631-82653-9 (MOBI)
DOI 10.3726/b17144

© Peter Lang GmbH
Internationaler Verlag der Wissenschaften
Berlin 2020
All rights reserved.

Peter Lang – Berlin · Bern · Bruxelles · New York · Oxford · Warszawa · Wien

All parts of this publication are protected by copyright. Any utilisation outside the strict limits of the copyright law, without the permission of the publisher, is forbidden and liable to prosecution. This applies in particular to reproductions, translations, microfilming, and storage and processing in electronic retrieval systems.

This publication has been peer reviewed.

www.peterlang.com

Contents

Introduction 11

1 The RESPECT period 21
    1.1 Life-giving rivers – balance between benefits and threats 21
    1.2 Waterways and bridges 23
    1.3 Cities and harbours 28
        Ancient harbours 28
        The development of inland and coastal navigation in Europe 30
    1.4 Military and defensive significance of rivers 32
    1.5 The structure of riverside cities 34
    1.6 Cultural connections 37
        Saint rivers 37
        The symbolism of water and river in culture and religion 41
    1.7 The milestones of ancient hydraulic engineering 42

2 The CONQUEST period 47
    2.1 Challenging water – the Dutch phenomenon 48
        Stages of polders reclamation 49
        Olęder settlement in Poland 52
    2.2 Navigation and industry – catalysts for urbanisation 55
        Transoceanic ventures – the colonial period 55
        The Industrial Revolution and development of waterway systems 57
    2.3 River engineering 59
        The Rhine waterway 60
        The Danube regulation in Vienna 63
    2.4 Water supply-sewage systems and pollution of rivers 68
    2.5 Canalisation of urban watercourses and drainage of swamplands 70

2.6 Beautification of the nature and cities	75
Parks and gardens	76
City densification	79
2.7 The scale of the changes and the price of the progress	83

3 The RETURN period ... 89

3.1 Raising awareness	90
3.2 Waterfronts revitalisation	101
Regaining the waterfronts of European cities	102
Waterfronts revitalisation in Poland	108
3.3 Environment regeneration	114
River regeneration at a regional scale	116
Reconstruction of urban watercourses potential	120
Blue-green networks	132
3.4 Water governance	136
Flood risk management strategies	137
Space for water in regions	139
Planning under conditions of uncertainty	143
Water quality	144
3.5 Integrated urban water management	145
A city as an ecosystem	145
Water governance in urban catchment areas	147
Blue-green infrastructure	149
Green architecture	151
Integration of water, environment and space management	154

4 Responsible cities – vital rivers ... 159

4.1 Responsibility as a way of return to rivers	159
4.2 Rotterdam – water challenges and assets	161
Harbour identity	163
Water threats in the time of climate change	168
Development strategies for the Water City	170

4.3 London and the Thames – friendship after divorce 173
       Water facilitated development 173
       Ecological cost of city development 175
       Revitalisation of the waterfronts and regeneration of London rivers ... 177
       Return strategies 182
  4.4 Singapore – "water wise city" 187
       Water "tiger" 187
       Water deficit 190
       The strategy of city-basin management 191
       Green infrastructure and architecture 194

5 Summary 199
  5.1 Return to friendship 199
  5.2 Vital rivers 206
  5.3 Urbanism and water 209

Abstract
River-friendly cities 213

List of figures 215

List of tables 223

Bibliography 225

Index of acronyms 251

Introduction

> *Don't divorce the river from its basin.*
>
> [H.B.N. Hynes, 1970]

Friendship with water?

Can a city befriend its river? Friendship is a cordial relation based on mutual kindness, trust and respect. Talking about friendship is feasible if it concerns at least two parties which respect each other's personalities. Cities, being human communities, are treated as entities in legal and economic terms whereas rivers lost their identity a long time ago. Urbanisation history has been a process of struggling with the nature whereas the level of technological subjugation of the element was an indicator of civilisation progress for centuries. Considering the scale of transformations of hydrographical structures in urbanised areas, the metaphor of "a river-friendly city" may seem to be an oxymoron like "urban planning for Gypsies"[1].

Relations between a city and its river tend to be judged "by the cover" i.e. by the quality of waterfronts. The cities which face their rivers, exposing the facades and bustling boulevards, attracting with the river bank lines, seem friendly and open towards the water. Entirely different impression is given by the areas turning their backs to the watercourses or being threatened by water therefore hidden behind embankments which protect them from the destructive power of the element. It is only one side of the picture, though. The other one is the rivers threatened by cities. Canalised, separated from their catchment areas, deprived of biocoenosis, contaminated, dead... What are the reasons behind the degradation of city watercourses? What processes made the cities divorce the rivers? Is their friendship still possible after the divorce?

RGB structures[2]

City environment is built for people by people. We have got used to perceiving its socio-economic, political-administrative, functional or

---

1 Umberto Eco, 1993, *Foucault's Pendulum*, Polish edition, Państwowy Instytut Wydawniczy, Warszawa, p. 80.
2 RGB is the acronym for Red – Green – Blue which refers to one of colour space models, determined by RGB colours but it is also used in spatial planning for defining urbanised

compositional-aesthetic aspects. In addition to the anthropogenic cultural structure, which I will graphically label as Red, a city also consists of natural (Green) and hydrographical (Blue) structures. All of them are packed into one limited space, interacting and constantly undergoing changes. Nearly by the end of the 20th century we had absolutely prioritised our Red structures, subordinating water and greenery to our own needs. "Natural assets" of soil, water, air and nature, which constituted the basis for cities' development, seemed so common and were hardly noticeable. The value and significance of services rendered by ecosystems began to be appreciated only when the consequences of the environmental degradation became an obstacle to further development and ecological footprint of big cities largely exceeded their surface area.

The analysis of city structures in RGB context casts light not only on the connections and interactions between the natural and built environment in a city or a region but also makes it possible to overcome the narrow anthropocentric approach to water which limits it to merely a source of benefits and threats for humans.

Urbanisation and water problems

The range of the anthropogenic environmental transformations, caused by urbanisation and industrialisation processes along with predatory exploitation of natural resources and greenhouse gas emission, resulted in ecological and climatic consequences. The occurrence of the phenomena has been recognised as Anthropocene[3], a new era in the history of the Earth, which features domination of human activity (Stoner and Melathopoulos 2015). In the 21st century the threats concerning shortage of water or its excess (floods), contamination and

---

(Red), natural (Green) and water structures (Blue). Its modified version called RGBG (Red – Green – Blue – Grey) also identifies infrastructural systems (Grey) – source Prof. D Arch. Galiano Garridos (University of Alicante), presentation: *The decision-making process for better city*, International conference RESPONSIBLE URBANISATION, Poznań 12.10.2018.

3   Anthropocene – the current geological epoch whose name was suggested by Paul Crutzen, a recipient of Nobel Prize in Chemistry (1995). The name refers to the influence of mankind on global natural processes, indicated by rapid urbanisation, irresponsible use of non-renewable fossil fuels, environment pollution and greenhouse gas emission (Steffen et al 2015).

destruction of ecosystems are becoming key issues of humanity and sources of international conflicts.

The World Economic Forum's Global Risks 2015 report placed water crisis at the list of the top global threats for the first time. The subsequent reports, besides weapon of mass destruction, also mention growing significance of global threats, including extreme weather phenomena, natural disasters or the failure of actions concerning adaptation to climatic change, which may even worsen the global water crisis (*World Economic Forum* 2018). According to the latest UN World Water Development Report (WWDR 2018), if the present level of water consumption continues, by 2025 two-thirds of the global population will live in water stressed areas. The prediction questions possibilities of achieving the 6th objective of the 2030 Agenda for Sustainable Development that is to "Ensure availability and sustainable management of water and sanitation for all" (UN 2015).

The scale of water problems increases along with cities population growth, especially in the middle- and low-income economies, where urbanisation pace is the highest.[4] Over 20-million cities like Bombay or São Paulo, whose population has surged almost tenfold for the last 50 years, have been struggling with basic problems of water supply and sewage treatment. The explosion of uncontrolled urbanisation (including growing slums areas) and its consequences have eventuated in threats for the inhabitants ranging from drinking water shortages and contagious diseases epidemics to flooding phenomena intensification (Kundzewicz and Kowalczak 2011). According to the UN projections, 68 % of the world population will inhabit urbanised areas by 2050. Vast majority will populate 10 million settlement structures of mega-cities. Already by 2030 as many as 43 cities will have reached mega-scale, while the fastest sprawling agglomerations will be the ones with fewer than one million inhabitants, mostly located in Asia and Africa (UN DESA 2018).

The above mentioned facts and predictions lead to a reflection: do urban planners realise their influence on the environment? What is the connection between city planning and water problems?

In the previous century, the fields of knowledge and expertise concerning water and space management underwent the processes of thorough separation and specialisation. As Shannon and Bruno De Meulder observe "pure urbanism",

---

[4] By 2050 the world urban population growth will be most distinct in India – 416 million growth, China – 255 million, Nigeria – 189 million (UN DESA 2018).

emerging at the end of the 19th century as a science, was devoid of determinants concerning water (2008, pp. 5–8). The issues beyond the interest and competence of urbanism included not only water supply and sewage treatment but also forming hydrographical structures as well as hydraulic and environmental consequences of city space transformations. No wonder, urban planners excused from considering water management, have ceased to take responsibility for urban catchment areas transformations. Also, architects forced to swiftly discharge rainfall water into combined sewer systems, did not feel the need to use water retention potential creatively. "An architect or a town planner did not have to think about water because a hydrology engineer prepared empty ground for new investments" (Nyka 2013, p. 117). "Cleaning" construction plots from greenery and draining urban swampland still prevails in Polish cities. The consequences are hardly noticeable at the scale of a plot, a quarter or even a district. We do not experience them within a year or even several years. However, if we consider a period of a century, 64 % of swampland areas[5] have disappeared globally and the process continues[6]. Only from 2009 to 2016 anthropopression resulted in disappearance of 33 % of the remaining water-swampland areas, and in Europe even as much as 45 % (Hu et al 2017). Urbanisation of increasingly bigger areas radically changes the conditions of water cycle and hydrological regime of watercourses. Whereas the consequences of neglecting the problem and failing to act at the scale of "your own yard", multiplied by millions of "yards", astound us with the extent of the disasters, severity of floods and intensity of urban heat islands. The hydrological cycle in cities is reduced to a system which directly discharges rainwater into "a pipe", leading to destructive flood problems at the "end of the pipe" in river valleys, seas and oceans.

The objectives, the plan and the scope of the book

The book aims not only to analyse the eco-hydrological consequences of urbanisation but also the process of transformations of city-river relations, generated by changes in values, views and development priorities in various historical

---

5   Source: GDOS.gov.pl <https://ochronaprzyrody.gdos.gov.pl/files/artykuly/5450/WWD18_Infographic_PL_icon.pdf> [accessed: 5.09.2018].
6   The review of 189 reports of change in wetland area finds that the reported long-term loss of natural wetlands averages between 54–57% but loss may have been as high as 87% since 1700 AD. There has been a much (3.7 times) faster rate of wetland loss during the 20th and early 21st centuries, with a loss of 64–71% of wetlands since 1900 AD (Davidson 2014).

periods. The changes show recurring nature of the developmental cycle. A question arises though: is it a spiral of advancement or regression? "In general, the current benefits tend to obscure the adverse effects, especially if the latter occur with a significant delay. The circularity of actions and consequences is inscribed in the nature of the world. The fact that we do not see it and do not understand is due to the lack of imagination and responsibility, as well as to the deficit of perspectives and cognitive tools" (Hausner and Kudłacz 2017, p. 205). The thought inspired me to survey European and Polish cities from a bit wider perspective in terms of space and time.

Analysing the state of the art, I considered the standards and stages of civilisation development concerning the transformations of city-water relations, interpreted by different researchers. In the European culture, since the beginnings of settlement there have been three principal development stages featuring diverse human approach to nature and perception of the role of water in the development of settlement structures. I outlined them as three periods: RESPECT, CONQUEST and RETURN in the subsequent chapters of the book. The purpose of the time travel was to describe the changes which otherwise remain unnoticed from the perspective of one generation.

The period of RESPECT (Chapter 1) was characterised by the attitude of both respect and fear towards water. The cities balanced on the edge of benefits and threats: on the one hand using transport, defence and production assets of water, on the other hand, respecting the flood risks. A distinctive feature of that period was also sustainable use of resources and subjective or even sacral approach to rivers, resulting not only from the peoples' dependence on water and lack of technical capabilities of resistance, but also from the holistic perception of the nature. Nevertheless, already in ancient civilisations the development of hydro-technical science and imperial aspirations of Rome opened the way for conquest.

The CONQUEST period (Chapter 2) was facilitated by cultural, political and technical transformations of modern era. The great geographical "discoveries", which introduced Europe into the era of colonialism, were a catalyst for economic growth based on exploitation of natural resources and non-Christian native cultures. The achievements in the field of reclamation and water engineering developed tools enabling transformation of catchment areas and river valleys. The Industrial Revolution period furthered the development of navigation, exploitation of rivers, industrialisation and rapid urbanisation which reached its apogee in the 20th century. The conquest appeared to come at the price of ecological disasters, floods, deterioration of living conditions and global climate change.

The period of RETURN (Chapter 3) followed a long process of raising awareness which led to changes in social values and city development priorities, manifesting themselves during the "green turning" of the 1970–80s. The stages of rebuilding relations with rivers included revitalisation of waterfront zones through regeneration of the environment and urban watercourses but also urban catchment areas renewal and integrated management of water, environment and city space. The discussed trends are illustrated with examples of numerous initiatives undertaken in European, American, Australian and Asian cities. The travel round "the global village" shows similarities between the problems and diversity of solutions within the scope of integration of water management with space planning, urbanism and architecture.

Chapter 4 "Responsible cities – vital rivers" discusses return strategies, formulated on the basis of the sustainable approach to anthropogenic (Red), natural (Green) and water (Blue) structures within city space, as well as restoration of cultural and eco-hydrological vitality of city watercourses. Examples of Rotterdam, London and Singapore illustrate different ways of return to friendship with rivers.

The summary (Chapter 5) includes the synthesis of the transformations in the periods of RESPECT, CONQUEST and RETURN, their technological-economic and cultural symptoms as well as eco-hydrological consequences. Comparative diagrams and tables allow to notice radical changes in urban proportions of RGB structures, the scale of the transformations as well as evolution of needs and pressure factors in the process of cultivating the relation between the city and water. Friendship requires respect and responsibility, therefore I conclude the book indicating possibilities of invigorating rivers in the cultural (R), natural (G) and hydrological (B) aspects and attempting to formulate a development direction for water responsible urbanism.

Relations with water as a civilisation development indicator – the state of the art

Relations between humans and the environment reflect socially accepted values which vary in different historical periods and cultural spheres. Evolution of socio-economic systems, built on the basis of culture, religion, science, technological progress and geo-political transformations, can be observed also in the structures of European cities. The approach to the environment, including water resources, determined the way of using rivers and developing the embankments, changes in water balance in city basins as well as priorities of space planning and water management in urbanised areas.

Historically, going back to Palaeolithic, settlement in the vicinity of water obviously stemmed from the necessity to meet the basic existential needs. Along with civilisation advancement the relations between the settlers and rivers strengthened, including not only defence, transport, sanitary or production functions but also gaining cultural-political, aesthetic and spiritual significance.

The first great ancient civilisations, whose cities were founded in the 3rd and the 2nd millennium BC, were born by huge rivers[7], which constituted the spine of settlement structures with their extent of waters and reclamation-irrigation systems. **Karl Wittfogel** labelled them as "hydraulic civilisations" (Wittfogel 1957), which expresses the major significance of water for the emergence of the cultures but also concerns the form of society structure which is necessary to build and maintain the irrigation systems and flood precautions. According to Wittfogel (1957) during that historical period only a despotic forms of governance, based on slavery, made it possible to control the water element with the use of primitive technology.

The level of navigation development was used by **Tadeusz Ocioszyński** as a means of civilisation progress assessment, which led him to establish three civilisation stages (Ocioszyński 1968, pp. 24–25): potamic (from Greek *potamos* – river), thalassic (from Greek *thalassa* – sea) and oceanic (Greek *okeanos* – the waters surrounding the globe), which featured diverse range of water space conquest. The first potamic civilisations lined navigable rivers but already the ancient cultures of the Mediterranean Basin, especially Phoenician, Greek and Roman ones, used seagoing fleet for founding colonies and developing trade. The connection between the civilisations and water showed in the shape of the harbour cities and transformations of the waterfronts. The progress in nautics, facilitating ocean crossings, initiated globalisation process and introduced European civilisation into the era of colonialism and spatial-cultural expansion.

Lewis Mumford's view (1934) may not directly concern the city-river relation but it reveals the changes in values and the influence of technology on civilisation development. On the basis of the concept by Patrick Geddes, Mumford indicated three distinct epochs: Eotechnic (Middle Ages), Paleotechnic (the period of Industrial Revolution) and Neotechnic (from the beginning of the 20th

---

7   Potamic civilisations flourished in fertile river valleys: Egyptian civilisation – by the Nile (approx. 3000 BC), Babylonian (approx. 3500 BC) – by the Euphrates and the Tiger, Indian (approx. 2500 BC) – along Indus, Ganges, and Brahmaputra Rivers, or Chinese (approx. 1500 BC) - from the "cradle" in the Huang-Ho basin to the entire area of the Yangtze and further to the south, while the mid-continental Iranian civilisation by the Amu Darya and the Syr Darya.

century to Mumford's contemporary times). According to the author (1934), industrialised, "machine" economy is the effect of our moral, economic and political choices. Mumford criticised uncontrolled expansion of cities and their financial-political structures (Mumford 1961), emphasising that "The physical design of cities and their economic functions are secondary to their relationship to the natural environment and the spiritual values of human community" (LeGates and Stout 1996, p. 91).

The natural and economic aspects of the relations between humans and rivers were analysed by **Krzysztof H. Wojciechowski** (Wojciechowski 2000), who observed the causes and directions for changes in three elementary periods: 1) original harmony of river functions, 2) domination of priority functions, initiated by the Industrial Revolution and 3) multi-functional economic use of rivers while raising awareness of detrimental effects of hydro-technical enterprises to the environment, simultaneously beginning to appreciate new economic river functions, especially recreational ones.

Similar periods were identified by **Marek Kosmala** (2011, pp. 5–6), who applied the criterion of relations with the environment as an indicator of civilisation progress. The changes, determined by humans' attitude to the nature, distinctively show in three periods: the stoical-sacral (from the beginning of settlement to the Industrial Revolution), subjugation-exploitation (from the beg. of the 19th century to the 1970s) and ecological-hedonistic one (initiated by U Thant's Report in 1969 and prevailing in the 21st century).

The development and transformations of waterside cities have been researched by many authors[8]. Publications on transformations of waterfronts worth mentioning include works by Rinio Bruttomesso (e.g. 1993, 1999, 2008, 2011), Ann Breen and Dick Rigby (1994, 1996) or Han Meyer (1999) who analysed the development of harbour cities. An invaluable source of knowledge on changing relations between humans and water can be found in a ten-volume publication series titled *Rzeki. Kultura, cywilizacja, historia* edited by Jerzy Kołtuniak (1992–2002). The landscape phenomenon of a city and its river, changing as time passes, was researched by i.a. Lucyna Nyka (2007, 2013), Alina Drapella-Hermansdorfer

---

8   Water (e.g. Tvedt, Jakobsson 2006), hydrology (e.g. Biswas 1978) and urbanism (e.g. Benevolo 1980; Morris 1994; Kostof 1991; Böhm 1994, 2006 and many others) are inseparably historically connected. More comprehensive analysis of the state of the art was included in my previous publications: *Water in urban public space. Model forms of rainwater and surface water development* (2011) and *Poznań Waterfront – Warta Valley. Revitalisation of the Relationship with the River* (2011).

(2004, 2005, 2009), Alina Pancewicz (2004), or Wojciech Kosiński (2009, 2011). Taking into account strengthening the relations between a city and its river as well as the character of water challenges, Lucyna Nyka (2013, p. 119) observes a new water urbanism trend consisting in negotiating the city structures which neighbour water bodies.

The developmental spiral cycle, mentioned by Jerzy Hausner and Michał Kudłacz (2017) when touching upon the concept of a city, can be also applied to the transformations of city-water relations which have been developing spirally from advancement to regress. Perceiving cities in multidimensional categories of Civitas and Urbs, allows to analyse the changes in physical space (spatial development), exchange space (socio-material relations) and discourse space (creating and sharing values) (Hausner and Kudłacz 2017).

Following the divisions suggested by K.H. Wojciechowski (2000) and M. Kosmala (2011), I identified three periods of civilisation development: RESPECT, CONQUEST and RETURN, determined by changes in humans' approach to the nature and significance of water in settlement structures. The names of the subsequent periods reflect the basic values and directions for transformations concerning city-water relations as well as their developmental spiral character.

Respect, combined with fear, accurately reflects the attitude of people to rivers in the early period of civilisation development. The respect resulted from the sense of entire dependence on water and admiration for its life-giving force but, simultaneously, the incomprehensibility and unpredictability of the element caused fear.

Conquest features aggressive expansion, confrontation, fight leading to the enemy's subordination and, ultimately, subjugation and exploitation. Such an attitude began to dominate in the relations between humans and rivers along with reinforcing the anthropocentric tenets and technological development.

Return does not mean heading towards the starting point but rather a change of direction at the spiral of civilisation development. The reality of the 21st century varies greatly and there is no possibility of return to the original balance in physical space; however, gradual return to respect of values and basic structures is a definite indication of desirable changes.

The time frame for the periods is difficult to be precisely determined. In some cultures (e.g. Maori), the "respect period" has lasted incessantly till today, in others the indications of the "conquest" appeared as early as in ancient times (e.g. in ancient Rome). In some cultural areas the "return" path has already been chosen, in others the exploitation and subjugation process still continues. Therefore, I limit my analysis to the European culture, occasionally referring to the achievements and influences of other cultures. Symptoms and key features of the transformations in the periods of respect, conquest and return are discussed along with their causes and consequences, considering the socio-economic-cultural (Red), environmental (Green) and water (Blue) aspects.

1 The RESPECT period

> No man made the land: it is the inheritance of the whole species .
>
> [John Stuart Mill, 1965]

1.1 Life-giving rivers – balance between benefits and threats

Ancient river civilisations were largely farming cultures, thus, the settlement was developed in accordance with irrigational river functions. High stages of river waters made it possible to water and enrich cultivated fields but they also threatened with floods. The settlement zones balanced on the line of high water stages which was determined by centuries-old experiences. The fertile areas by the river banks were used for crops while the solid buildings tended to be located above the flood plain terrains.

Water abundance and river alluvium provided excellent crops. In the subtropical desert climate of Egypt or Mesopotamia vegetation conditions deteriorated along with the growing distance from the river valleys, therefore, the agriculture was based on vast irrigation systems, whose best known examples were constructed in Egypt. The rising Nile could deeply penetrate the cultivated fields owing to the irrigation canal systems while the application of simple bulkheads and tools for drawing water, such as Archimedes' screw[9], made it possible to distribute it to the higher located areas. The regularity of the overflows and the competence in measuring their level allowed to precisely predict the amount of the crops which gave the Egyptian clergy a tool of political power.[10]

River valleys and estuaries were particularly susceptible to floods. Already at the beginning of the 2nd millennium BC in Egypt, the first canals and storage reservoirs[11] were built which partially allowed to protect the

---

9   The invention attributed to Archimedes, who lived in the 3rd century BC, is actually a Babylonian invention also used in ancient Egypt.
10 Nilometres – measuring structures in the form of an underground corridor with a well at its end, which reached ground waters and were connected with the temples. Cf. Arnold D., *Die Temel Ägyptens*... 1992, pp. 41–42.
11 Draining and retaining excess waters of the Nile allowed to develop the great Faiyum Oasis (Stanielewicz 1995, p. 156)

delta[12] from river flooding or sea storms. The ancient potamic civilisations constructed water dams, the oldest of which were built in Mesopotamia, in order to control the flow of the Euphrates and the Tigris[13]. In India in approx. 4th century BC, water measurement and charge calculation systems were commonly used but also dams[14] appeared along with water retention canals and reservoirs[15].

The large rivers were uncontrollable though. The Chinese Huang-ho river changed its course and the estuary place several times, causing changes in the settlement system. Similarly, the Tigris and the Euphrates often changed their courses, moving away from the cities and exposing them to annihilation in the desert sand. Jerzy Makowski (1997, pp. 197–225), referring to an expert on the subject, W. Müller, labels Babylon as "Holland of antiquity" as its inhabitants were forced to build dams, canals and earthwork along the banks of the capricious rivers[16] in order to protect their settlements and cultivated fields form floods. The higher the flood revetments were, the more disastrous the floods appeared, managing to destroy the structures. Probably one of such tragic floods became the basis of the myth about the deluge and Noah's Ark (Orłowski 1993, pp. 41–42). In Babylon, the gigantic floods led to archaic fear of water as the gods' punishment. However, the dread and humbleness caused by the element did not restrain water engineering progress.

---

12 The author of the first description of the settlement in the Nile delta was a Greek historiographer Herodotus from Halicarnassus (485–425 BC) who, examining the multi-branch Nile delta, named it Δ (delta) – the fourth letter of the Greek alphabet.

13 Java in Jordan is the oldest known water dam, dating back to 3000 BC. The gravitational structure was a stony wall, 9 m high and 1 m wide. It was supported by 50 m wide earthwork. Sadd el-Kafara at Wadi Al-Garawi, 25 km from Cairo, is another example. The structure is 102 m long and 87 m wide and was built in approx. 2800 BC (Garbrecht 1986, pp. 51–64). Quatinah Barrage in contemporary Syria is considered the oldest still existing dam. Its development dates back to the time of pharaoh Sethi's reign. (1319–1304 BC).

14 In this way Sudarshana Lake in Girnar was created. Cf.: Panikkar K 1967, p. 43.

15 In India, due to seasonal rainfall and considerable fluctuation in water level, a special type of very deep (20–30 m), stair-shaped wells was developed. They were multilevel containers in the form of earth-sheltered cloisters (e.g. Lalkot in Delhi). "The stairs or multilevel arcades lead down to the water surface, being covered or uncovered by changing water level. Huge stairs, stepping down to the water surface, became one of the particularly unique features of Indian architecture" (E. Niemczyk 1995, p. 85).

16 More: Prof. Ph. D Paweł Jokiel, 2011, „Mokre" konflikty. Pracownia Hydrologii i Gospodarki Wodnej Uniwersytetu Łódzkiego < http://hydro.geo.uni.lodz.pl/index.php?page=mokre-konflikty> [accessed: 6.07.2018].

The extent of river flooding determined a strict boundary for construction of permanent city development. In close river vicinity, harbour facilities and craft workshops were constructed for the water use in technological processes; nevertheless, they were temporary and low-budget buildings due to their possible destruction by flooding waters. Rivers were unpredictable whereas chaotic embankments, dykes and dams frequently caused obstructions and catastrophic overflows. The benefits from locating cities in the fertile river valleys and deltas in the immediate vicinity of waterways frequently came at the expense of damage triggered by floods. The settlers faced then the dilemma of moving the settlements to the higher located areas[17] or building embankments and dams.

In the medieval Europe, partial river embankments[18] were built mainly to protect the arable land as the castle towns and cities were located above the flood plains. "The structures were chaotic for every ground owner attempted to save his land, not taking into account their neighbours or the river changes after building the embankment. The actions brought more damage than benefit. Growing population along with the valley development led to increasing process of narrowing the big river bed. Consequently, the damages caused by the floods soared" (Bobiński and Żelaziński 2015).

The denser the settlement system became, the more severe damage was caused by floods. Nevertheless, in Europe there were still vast flood plains, swamps and wetlands which mitigated the impact of the violent overflows.

1.2 Waterways and bridges

Rivers provided abundance of water and food and "as the settlement stabilised, the tools improved, the social division of labour increased and, consequently, the exchange of produce surplus spread, the transport function of the natural waterways acquired significance"(Gan 1978). In the age when horses and camels had not been domesticated yet, but also in later periods, deserts, backwoods and swamplands posed an obstacle which was difficult to overcome by people and pack animals. Whereas using the waterways saved the effort of penetrating the wilderness, searching for fords and facing dangers lurking on land,

---

17 J. Makowski (1997, p. 197) gives Polish examples of Kwidzyn (beg. of the 14th century) and Świecie (mid-19th century) moved for the flood threat.
18 In Western Europe in the 12th century, the embankment of the Rhine began, while Henry II Plantagenet initiated the embankment of the Anjou Valley (the lower Loire) (Makowski J., 1997, p. 199). In Poland, the construction of the embankment in the delta of the Vistula river began in 14th century under the reign of Casimir III the Great.

which made water communication far more convenient, faster and safer. Despite obstructions, shoals, knick points and seasonal problems resulting from freezing or drying, rivers were nevertheless the ready-for-exploitation trails, requiring only the adjustment of forms and sizes of boats to the watercourse type. Natural water movement made the journey along the river course require minimal physical effort (Gan 1978) while the open valley space allowed to spot the threats from a long distance.

In ancient Egypt the Nile course "facilitated shipping in the Mediterranean Sea direction while usually northern wind powered the sails of the ships sailing up the Nile. The irrigational canals enabled the ships to sail transversely against the Nile valley and the dense net of the Nile delta tributaries also furthered shipping distribution. As a result, the wheeled vehicles had been unknown in Egypt by 1700 BC as they were redundant" (Piskozub, 1993). The water transport on the Nile made it possible to build pyramids, complexes of temples and first cities located along its banks. The stone building material was delivered to the building site by boats owing to transport canals, specially dug just for that purpose. The Egyptians also built huge navigation canals e.g. in approx. 2150 BC in the area of Asuan, a canal was built enabling big ships to overcome the first Cataract of the Nile; and during Necho's reign (610–595 BC) the Egyptians attempted to build an antique version of the Suez Canal, whose northern section connected Great Bitter Lake with the Nile (Stanielewicz 1995, p. 159).

In China, the main transportation route was on the Yangtze river with its tributaries and canals while the land roads were used solely for communication. The inland navigation was slightly less important in India and Mesopotamia due to different direction and profile of the rivers (Piskozub 1993) but already during Hammurabi's reign (1728–1686 BC)[19] the Euphrates and the Tigris were used to raft wood, cereal, olive, wool, date fruit, leather and handicrafts (Gan 1978).

Communication-transport water arteries constituted the skeleton of settlement structures and a catalyst for the development of cities. Harbours, waterfronts and navigation canals were adjusted to the types of ships and boats. Erosion-sedimentation processes constantly changed navigation conditions therefore river ships were usually limited in size due to the necessary adjustment to natural waterways[20].

---

19 *Code of Hammurabi* gives numerous examples of details concerning Babylonian navigation.
20 Average Nile ships (according to records from the 15th century BC) were narrow and long (up to 20–22 m), and sea ships of Egyptian or Phoenician type, were 6 m wide and 50 m long (Stanielewicz 1995, p. 156).

Rivers were a real obstacle to the land road traffic. Primeval people overcame water barriers in various ways, most frequently they swam, used tree trunks, later rafts and boats as well as wading places and bridges. During the reign of Cyrus the Great, Persians began to build pontoon bridges, which floated on the inflatable leather water bags or boats[21] but the breakthrough significance lay in the construction of bridges above water – permanent structures enabling the connection between two banks of the river. The first solid big-breadth bridge on stone pillars was built in Babylon over the Euphrates already in approx. 620 BC (Stanielewicz 1995, p. 158).

The Romans appeared to be masters at building bridges thanks to the application of stone arches[22] and stable installation of bridge counterforts on ground[23] or on alder stilts (Stanielewicz 1995, pp. 158–160). In Renaissance the Roman methods of constructing bridges were still followed and improved. In view of its strategic and economic significance, "bridge" gained importance as a symbol of unification and connection in defiance of the dividing power of the river[24].

Bridges constituted nodal points in urban space – strategic elements of communication facilities which enabled the development of city parts separated by its river but they also created bottlenecks – places where land communication was channelled to a narrow river crossing. It affected the riverside cities layout which often reminded an hourglass (with the narrowing in the bridge crossings) while the roads leading to the place became the skeleton of the spatial city structure. Prague by the Vltava (Fig. 1.1), Paris by the Seine (Fig. 1.4), London by the Thames, Toulouse by the Garonne or Florence by the tiny Arno river are finest examples of cities located on two banks of their rivers and connected by bridges.

A wide river technically impeded cities development thus, for instance Vienna by the Danube (Fig. 1.2), Koln by the Rhine (Fig. 1.3), Bordeaux or Tours (France) developed only on one side of the river for a long time. In Poland, such

---

21 During the war against Greece in 480 BC, the Persians developed 1 200 m bridge over the Dardanelles, floating on 300 boats.
22 The arch spans of Roman bridges reached 25 m, and their height – 48 m. "The biggest technological achievement of building river bridges was the bridge dating back to 105 AD, built over the fast flowing Danube by Iron Gates, during Emperor Trajan's reign. 1 100 m long and from 13 to 19 m wide, it soared 46 m over the water surface and was supported by 14 huge pillars" (Stanielewicz 1995, p. 160).
23 The Romans used hermetic bulkheads of wooden stilts for this purpose.
24 In Muslim culture, a bridge (*al Sirat*) was the connection between the earth and the paradise which could be only crossed by religious people (Łagiewski 1993, p. 169).

Fig. 1.1: Prague, 1608, Abraham Saur, <https://www.vintage-maps.com/> [accessed: 18.11.2018]

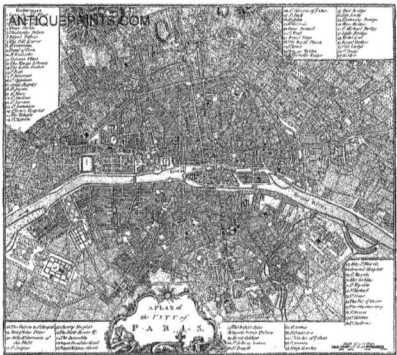

Fig. 1.4: A plan of Paris, 1776/1800, John Andrews, <http://www.ancestryimages.com/> [accessed: 18.11.2018]

cities as Szczecin, Toruń, Płock, Racibórz or Opole extended only one-bank structures in view of the technical limitations concerning long span bridges.

Urban bridges were such important pedestrian communication routes that they were transformed into covered trade arcades like Ponte Vecchio in Florence or Ponte Rialto in Venice. Owing to their architectonic forms and wide display foreground in river valleys, they have been landmarks of urban riverside

Fig. 1.2: Vienna, 1608, Abraham Saur, <https://www.vintage-maps.com/> [accessed: 18.11.2018]

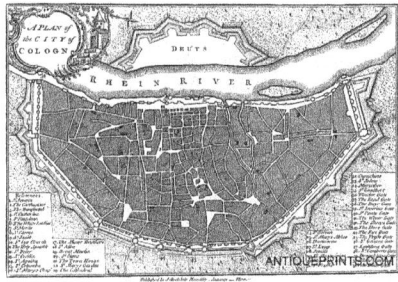

Fig. 1.3: A plan of Koln, 1800, John Andrews, <http://www.ancestryimages.com/> [accessed: 18.11.2018]

landscape[25] and city panorama viewpoints. In Prague, which was one of the biggest cities in Central Europe during the reign of Charles IV (1355), the river was deliberately used to create a representative city panorama (Fig. 1.1) while Charles Bridge over the Vltava, constructed in 1357, became the main axis of the city layout. The Vltava, emptying into the Elbe north of Prague, was then part of a huge water route, connecting the north of Central Europe with its south.

---

25 More i.a.: Brown 1996; Brown 2005; Flaga et al. 2005; Bonenberg 2007.

1.3 Cities and harbours

Ancient harbours

The crossings of water and land routes created particularly convenient conditions for settlement advancement. Babylon, one of the greatest cities of Babylonian civilisation[26], was the capital of Babylonia, a country developed in the 3rd millennium BC in Mesopotamia (currently the area of Iraq)[27]. The lively, multicultural city, famous for its palaces, temples and huge ramparts, grew rich owing to trade routes and craft development. Alexander the Great, appreciating the trade and military significance of river and seagoing fleet, planned to establish a harbour for one thousand ships in Babylon and brought Phoenician sailors there (Montesquieu 1927, p. 176). Shipping flourished not only on the Euphrates and the Tigris. In the 5th century BC, Herodotus wrote that the entire Babylon was cut with canals and the biggest of them, in the south-east, was navigable and stretched from the Euphrates to the Tigris. The information concerned the Royal Canal which systematically deepened and reconstructed had still functioned by the 7th century AD (Gan 1978).

The Harappan culture developed in the Indus River valley and its basin. The main centres of the culture were located in the cities: Harappa and Mohenjo-Daro. Mohenjo-Daro, located on the top of a hill in the Indus floodplain, distinguished itself by its vast water supply-sewage system (cf. Chap. 1.7).

Already at the turn of the 3rd century BC the first European culture centre and marine coastal navigation developed[28] in Crete, enabling the island and Greek countries to contact. The Palace of Knossos, located merely 4 km from the sea shore, was connected with the harbour by the road and the river.

The Greeks were not only excellent sailors but became famous for their skills in constructing harbours. Short mountain rivers of Hellas did not have trade, strategic or communication significance and could not compete with the sea in

---

26 The first settlements in Mesopotamia were developed by the Sumerians approx. 3500 BC. Each of them consisted of a city surrounded by smaller settlements.
27 Even earlier, Uruk existed as a city already in the early Uruk period (4000–3500 BC). It was then one large city in the southern Mesopotamia (400 ha).
28 The Phoenicians, who first could build wet docks with piers and wave breakers, were the forerunners of oceangoing shipping in antiquity. Tyre and Sidon, the main Phoenician harbours, were the basis of their trade empire by the Mediterranean Sea. The colony founded by them in Carthage (ca. 814 BC) with two regularly shaped harbours (the circular martial one and the rectangular trade one – Ostrowski 2001), became a power which was destroyed by the Romans as late as in 146 BC.

this respect, therefore, most of the Greek cities owe their advancement to the convenient location of the sea harbours. The Greek thalassocracy[29] developed in the Balkan Peninsula and numerous islands of the Aegean Sea and soon spread to the entire Mediterranean and the Black Seas basins. The oldest Greek harbours discovered on the islands of Delos (from the 8th century BC) and Samos (540–523 BC) already had grand seafronts and piers.

Athens flourished thanks to its location in the vicinity of one of the most convenient natural harbours in Greece. Wide loading berth in **Piraeus** had communicational connection with agora. **Miletus** evolved on the headland between two bays, where the integration of the city layout with the harbour location allowed to create representative water fronts and important city public space in the Lion Harbour. In Greek cities (e.g. Knidos) one can observe not only the adjustment of the harbour location to the natural environment but also the compositional integration between the city plan and the shape of the harbour. Practically, the majority of harbours were a kind of the city outwork, partially separated for the risk of an attack from the sea side or the danger of a storm.

Alexandria, established in 332 BC by Alexander the Great[30], advanced thanks to its strategic location in the Nile delta, by the Mediterranean Sea and Mariout Lake while its connection with the arm of the Nile made it possible to integrate the river and sea transport. The sea route to the Magnus Port, between the city and Pharos Island, was lit by a huge 12–140 m lighthouse, built in 2nd century BC, which was declared one of the wonders of the Ancient World. According to Wacław Ostrowski (2001), foundation of Alexandria was "the first symptom of the end of the great river valley period and increase in the importance of the Inland Sea. [...] City settlement of the Mediterranean region also spread to another inland sea – the Black Sea which became surrounded by Greek cities. **Byzantium**, established by the Greek colonists in the 7th century BC at the crossing of the water route (connecting two sea basins) and land routes leading from Asia Minor to Europe, gradually dominated the others especially when Constantine the Great moved the capital of the Roman Empire there in 330. The city, constantly extended and beautified, became one of the main centres of European culture" (Ostrowski 2001, pp. 175–176).

---

29 In accordance with the division by Ocioszyński, the ancient Greece and Rome belong to thalassocratic civilisations (Ocioszyński 1968).
30 Alexander the Great founded Alexandria on the ruins of the Egyptian settlement of Rhacotis from the Old Kingdom period.

The Roman Empire had an extensive land road system, mainly used to relocate troops; however, goods transport, especially cereal from remote colonies in Gaul and Egypt, required the development of water routes and harbours. The sea harbour in **Ostia**, from which cereal was transported to Rome using the Tiber water route, was a window to the world for the capital of the empire with its sea lighthouse modelled on the Alexandrian one. The harbour cities not only became buzzing trade centres but also the display of power and the Roman builders' technological skills.

The glamorous structures of harbours from Hellenistic times in Carthage, Rhodes or Alexandria, were monumental gates and the showcases of the cities. The structures placed in the harbours such as the gigantic statue of Helios on the Rhodes Island[31] or the sea lighthouse on Pharos, in addition to their navigational significance, also had the symbolic meaning of a challenge to the great water and the announcement of the conquest era.

The development of inland and coastal navigation in Europe

The inland and coastal shipping which developed in Europe in the twilight of the Middle Ages became the catalyst for cities development as water transport was a convenient, cheap and relatively safe means of transport. The harbour cities had especially favourable conditions for development, being located on the crossings of sea and inland waterways and land routes.

Beginning as early as in the 12th century, irrigation-transportation canals were built in the Northern (in the Netherland) and Southern Europe (in Lombardy)[32]. **Milan** owes the development opportunities to the construction of **Naviglio Grande** (1179–1233), whose modernisation in the 15th century was guided by Leonardo da Vinci[33]. He was also the author of multi-space hydraulic

---

31 The Colossus of Rhodes (32–36 m high, built in 304–292 BC) and the sea lighthouse on Pharos (built in 280–279 BC) became two of Seven Wonders of the Ancient World.
32 The construction of the first 30 km long section of the Naviglio Grande canal lasted for over 50 years (1179–1233). After its extension in 1258, the canal reached Milan. Source: *Od przewłoki do podnośni. Jak na przestrzeni wieków ewoluowały urządzenia i budowle hydrotechniczne ułatwiające żeglugę i umożliwiające tworzenie sieci wodnych dróg śródlądowych*<http://kanalgliwicki.net/sluzy/index.html> [accessed: 15.07.2018]
33 In Atlantic Codex, the notes by Leonardo da Vinci from 1489–1492, there is a drawing of an innovative solution of closing the lock, called miter lock, with the use of a supporting gates which seal when pressed by water. San Marco lock (1497), joining Naviglio Grande and Martesana canals, was the first lock equipped with such kind of gates. Five other locks on the Naviglia Grande were also attributed to Leonardo da Vinci. Op. cit.

engineering plans[34]. Advances in hydraulic and water engineering gradually paved the way for conquest.

Since the 12th century, following the collapse of famous ancient sea harbours: Ostia, Pozzuoli, Brindisi, the trade in the Mediterranean Sea basin was taken over by three cities: **Venice, Pisa, Genoa**, which owed their development to their harbour and trade functions as well as the excellent fleet. Pisa, lining the Arno river, became a significant trade centre after Augustus had built the harbour. Those cities-states can be labelled as thalassic cultures of the Middle Ages and Renaissance[35] (Benevolo 1995, p. 37–51). Similarly, in the Northern Europe, **Bruges**[36] and later **Antwerp**[37] enjoyed primacy.

The development was also furthered in the 13th century by the foundation of the Hanseatic League, a mercantile-military confederation, which later transformed into confederation of seaside cities of the North Sea basin (Benevolo 1995, p. 75). In the 15th century, the period of the biggest prosperity, the Hanseatic League comprised ca. 160 coastal and inland cities from different countries, including Hamburg, Dinant, Göttingen, Halle, Kalmar, Reval, and Polish cities: Wrocław, Kraków, Gdańsk, Chełmno, Elbląg and Toruń.

---

34 Leonardo da Vinci suggested i.a. the drainage of The Pontine Marshes near Rome and the construction of a canal bypassing the Arno River in order to connect Florence with Postojna as well as the irrigation of the Pad River valley and creating the canal system connecting Milan with the southern Italian lakes (Böhm 2006, p. 31).

35 Sea voyages and land travels (e.g. Marco Polo 1271–92) as well as the Venetians' and the Genoese's geographical discoveries contributed to the development of shipping and technology and implementation of other cultures' inventions in Europe. It is plausible that the Chinese invention of the lock chamber came to Europe through the Arab merchants or owing to Marco Polo's expeditions.

36 Burges – the biggest merchant city of Europe beyond the Alps, was founded at the end of the 9th century around the fortified castle by the Reye river, near the sea bay which goes deeply inland and has excellent sailing conditions. A trade settlement developed on its shore. In 1134, as a result of a storm, a deep bay – the Zwin – appeared at the end of the former arm of the river, where the new outer port, Damme, was constructed and connected with the Reye river and the city by a canal. Owing to such a location, Burges became the main harbour of Europe in the North Sea (Benevolo 1995, pp. 72–73).

37 Antwerp, located by the Scheldt estuary flowing into the North Sea, became the most important and the richest trade city of Europe in the 16th century. Convenient communication conditions facilitated trade development, which resulted in the foundation of the first stock market in the world in 1531.

1.4 Military and defensive significance of rivers

The riverside location of settlements made it possible to use and control water routes as well as fleet movements, while the wading places and bridges allowed to control the land communication tracks. The capitals of Roman provinces: Trier, Milan, Sirmium and Nicomedia were strategically located by the great rivers but also small rivers, wetlands and backwaters were used for enhancing defensive capability of settlements, expanding water structures with artificial canals and moats.

The strategic importance of water courses is proved by the fact that the borders of countries were mostly marked along large rivers, which were natural obstacles to tribes' expansion and constituted warfare lines[38] (Herman 1998). In the Middle Ages, two European rivers: **the Rhine and the Volga** had great military significance owing to their size and border location (Margul 1995, pp. 61–62). The defensive profile of the Rhine shows in the line of fortified cities-camps established by the Romans, who developed so called limes[39], linear fortification facilities along the rivers. The Rhine separated the grounds of the German tribes from the Roman empire land and, by the 2nd century AD, it had constituted a firm defensive strip of so called ***Limes Germanicus*** with dense location of fortified cities on the left bank of the Rhine (Basel, Strasbourg, Speyer, Worms, Mainz, Koblenz, Bonn. Koln). Also, **the Danube** was defended by fortified buildings or land plots of Roman legions[40]. **The Volga**, the biggest river of Russia, was the eastern frontier which protected the Slavic people from the Asian tribes. While the defence line of the Dniester, with Serpent's Walls sprawling in the areas of Ukraine, Moldavia and Romania, was supposed to shield the Slavic people from the invaders from the south. At a local scale, smaller rivers played a similar role, e.g. the Bóbr river in Lower Silesia whose middle- and down-stream were included in the defensive line of Slavian Chrobry Embankment.

---

38 The defensive importance of rivers was used by Alexander the Great as well, by marking out the borders with Persia and India along the riverine frontiers.
39 The Roman writers presented limes as sui generis "holy border" which people did not cross. If they did, it was treated like overstepping the boundary of common sense and civilisation. Source: Imperium Romanum <https://www.imperiumromanum.edu.pl/geografia/limes/> [accessed: 22.07.2018].
40 The Romans established i.a. Aquincum which developed into Budapest. Also, in the 15th BC Vienna was a Roman border point (Vindobona), guarding the Roman Empire from the northern German tribes' invasion.

At the early stages of civilisation development, the most important decisions concerning the location of a settlement had to allow for the defensive possibilities. Both in the case of the settlements which were developed spontaneously and the cities founded in accordance with a plan, the defensive advantages of rivers were used in order to hinder the access for prospective invaders. In the late empire period, the Roman cities were surrounded by walls and also natural water obstacles including rivers, backwaters, swamplands, steep banks and even parts of aqueducts (in Rome).

The fact that the Romans considered **the Tiber** as a border was confirmed by the plans of Julius Caesar who intended to move the riverbed further to the west in order to incorporate new territories (Herman 1998). After the division of the Roman Empire, the capital of the western empire was moved to Ravenna for its protective location by the very sea as well as inaccessible swamplands and back waters. The city was well defended and, simultaneously, it had a waterway connection with the empire owing to the vicinity of the Classe harbour, dating back to Augustus' reign. Rome, having lost the water supplies from the aqueducts as well as the cereal consignments from Ostia harbour and food stores along the Tiber, collapsed after the Goths invasions (535–553 AD) while the survivors found shelter in the river backwaters. In this way, despite the fact that the Tiber threatened with constant overflows, it became the cradle of the new Rome (Benevolo 1995, pp. 23–31). Former Roman camps, transformed into cities surrounded by walls but lacking the support of the powerful army, used the defensive river lines, turning into a system of distant fortresses.

Janusz Bogdanowski (2000, p. 204) identified eight types of connections between the riverine location and defensive capability: 1) "a fortified place" – a sole spot by the river, 2) "a gate" on the border of the land or the whole country, 3) a settlement "area", 4) a strategic "zone", especially along the river estuary, 5) "a bridgehead", 6) "land" surrounded by river backwaters, 7) "a border zone" protected by wetland valleys of rivers, 8) a river confluence fortified like a "manoeuvre polygon".

In terms of defensive capability, the settlements **on islands** (e.g. Saint Michael's Mount, Venice, Stockholm) were protected in the best way. For that reason the Medieval castles of defensive character began to be surrounded by water canals (**moats**), which hindered the access to the walls. The defensive capability of Polish and Teutonic Order castles, built since the 14th century (e.g. Olsztyn, Malbork, Toruń, Świecie, Chełmno), was skilfully enhanced by riverside location. The cities situated in the **confluence of rivers** (e.g. Paris, London or Prague) were protected by water from a few sides. Founded on both banks of a big river, they were in fact two independently fortified settlements or a city and a beachhead on the other side of the river. Such kind of location gave superb control over the water routes. The limitations resulting from the choice of hardly approachable riverside locations can be observed in cities layout and in overcoming subsequent developmental stages.

Medieval moats, and later Renaissance bastion forts with water foreground, became a standard of building cities in the entire Europe. Created at the turn of the 15th century in Italy, radial layout of ideal cities was not adjusted to natural topographical conditions but rather followed geometrical water-space arrangements whose construction was enabled by the progress of water engineering in the 16th and the 17th centuries. Artificial canal systems, supplied with river water, survived in cities as long as until mid-19th century; though, along with the progress in military technologies, the defensive significance of water diminished while the redundant and polluted canals and moats were gradually backfilled.

1.5 The structure of riverside cities

Urban structure of Europe was created from the second half of the 11th century to the second half of the 14th century. Tadeusz Tołowiński (1948) defined six main city-creating factors: 1. natural conditions, 2. socio-economic, 3. defensive, 4. communication, 5. custom-legal, 6. urban layout. Water plays major role in all the aspects. Anthony E.J. Morris (1980) pointed out only two groups of factors: natural and anthropogenic. Rivers were one of key topographical determinants of urban form. Original dependencies between hydrological system and urban form resulted mainly from **transport and defence** needs. Stanisław Liszewski (1995, p. 133–134), referring to former research by M. Kiełczewska-Zaleska (1969) and W. Różycka (1965), defines a few types of Polish medieval cities location which take into account the defensive advantages of the river:

- insular (e.g. Wrocław, Poznań, Głogów) – location on an island, created between anabranches, impeded the access and made it possible to observe the broad river valley as well as control the wading places which, as a rule, emerged in the shallows of backwater.
- valley-bottom (Racibórz, Koźle, Gdańsk) – cities located in the valley, above the floodplain, were convenient for the river and had quite good defensive advantages but were threatened with floods during river overflows.
- edge (Grudziądz, Włocławek, Toruń, Puławy, Warszawa) – location on the high river bank, and particularly on the embankment (Kraków, Płock), prevented from flood and, simultaneously, facilitated defence against enemies.

The influence of topographical conditions on riverside cities structure was also analysed by Alina Panacewicz (2004, Table 2, p. 54), who identified nine specific forms of location of historical settlement complexes by rivers.

It seems functional connections between city tissue and water as well as representative waterfronts evolved as a result of the **advantage of communication functions over defensive** ones. Along with economic boom of late Middle Ages, the significance of inland and coastal navigation grew, as it became the foundation of transport system of the contemporary Europe. The most distinctive element of a Medieval city plan, representative for the North Sea and the Baltic Sea basin, was a **comb-shaped street layout** determined by harbour functions.

Lübeck, one of the first harbour cities on the Baltic coast founded by Henry the Lion in the second half of the 12th century, had superb defensive capabilities, guaranteed by water courses of the Trave and the Wakenitz as well as vast swampland in the east. The spine of the city consisted of a trade route from the south to the north and a harbour by the Trave river connected with a trade centre by the comb-shaped street layout (Fig. 1.5). Lübeck owes its development and the specific space layout to the Hanseatic League membership. "Lübeck became a top member of the Hanseatic League […] which owned numerous exchange offices or even whole districts in London, Bruges, Bergen, Nowogród and other international trade centres. Many cities adopted the Lübeck law; they used comb-shaped layout of the streets leading to the harbour. […] The spatial structure and plan of Lübeck provide an example of the mature concept of flourishing Middle Ages, considering both functions and social relations but also special qualities of the natural environment" (Ostrowski 2001, p. 181).

The comb-shaped street layout, which was used for the first time in Lübeck as a compositional arrangement resulting from close communicational and functional connections between the harbour embankment and the trade city centre, became a pattern for many harbour cities of Medieval Europe. Founded in 1255 Stockholm, located on an island, had two-side comb-shaped street layout, connecting the waterfront with the trade centre while the access from the land (dyke) was protected by a fortified castle.

In Poland, similar street systems were constructed in cities founded by Teutonic Order such as **Toruń**, **Gdańsk** (Fig. 1.6) and **Elbląg**. "The comb-shaped layout, dating back to the beginning of the 14th century is visible in the Main City of Gdańsk. Nine streets led to the Motława river waterfront; almost all of them were provided with loading gates and bridges by the river. The largest gate, called the Crane, was used for defence and transhipment. Its huge structure constituted a meaningful sign of Gdańsk role in European trade" (Ostrowski 2001, pp. 181–182).

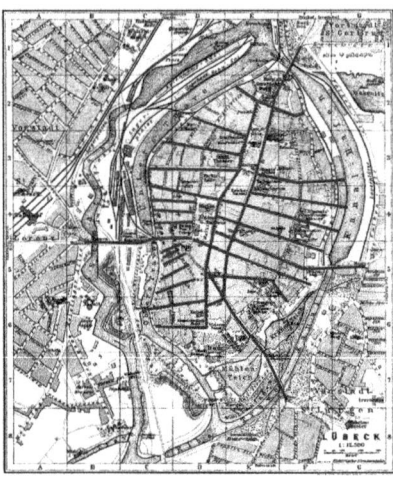

Fig. 1.5: Lübeck, 1910. Own study based on: <http://www.lib.utexas.edu/maps/map_sites/hist_sites.html> – Courtesy of the University of Texas Libraries, The University of Texas at Austin

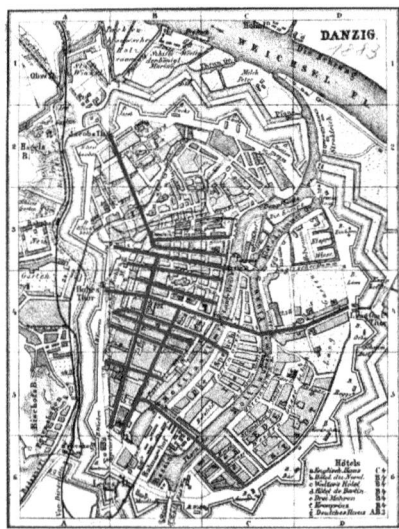

Fig. 1.6: Gdańsk, 1883. Own study based on: UMGEBUNG VON DANZIG, Baedeker Karl, Mittel- und Nord-Deutschland, ed. GeographischeAnstalt von Wagner & Debes, Leipzig 1883.

Other Polish cities like Brzeg by the Odra or Warsaw by the Vistula river, not belonging to the Hanseatic League, did not develop structural connections with their rivers, despite great significance of river transport. Founded on a high riverside scarp, in accordance with the plan based on orthogonal street system with a rectangular market, they were not directly linked with harbour berths. Only increased number of urban doors and gates as well as roads leading to births and riverside-situated mills, fulleries, butcheries, tanneries and other craft facilities, show the functional connections with the river. Advantageous location on one riverbank or slightly further from water, like in the case of Vienna or Warsaw, enabled the cities to retain merely transformed sections of river valleys. Wildlife enclaves (e.g. the right Vistula river bank in Warsaw, the left one in Toruń, the Dębina in Poznań) which were considered abandoned and undeveloped areas for years, now appear to be invaluable natural heritage and showcases of the 21st-century cities.

Antonio Averlino, called Filarete (1400–1469), the author of the first thesis on constructing cities (approx. 1460), described an example of the ideal city of *Sforzinda,* with its eight-point star layout and optimal location in a fertile river valley, surrounded by hills. The river was supposed to flow near the city, enabling sewage discharge, but was not to flow into the city in order to avoid floods (Ostrowski 1966, p. 395). Also Leonardo da Vinci suggested creating canal systems to remove sewage, anti-flood protection facilities and extension of waterways.

1.6 Cultural connections

Saint rivers

Considering the contrast between a green oasis supplied by streams and the sunburnt, hostile desert, it is not difficult to comprehend the worship of the life-giving power of water in Mesopotamian or Egyptian civilisations. The Biblical paradise consisted of four rivers which transformed the desert into blossoming gardens. The Book of Genesis (Gen. 2,10–14) mentions the river systems of the Nile, the Tigris-Euphrates as well as the Indus as the four tributaries of the paradise river. Four rivers of life also appear in the Iranian cosmology[41] while the gardens of

---

41 According to Iranian cosmology, the Earth developed from water whose goddess was Aredvi Sura Anahita having the power of purification and life giving fertility.

Persia, whose patterns were shaped from 800–400 BC, were the Zoroastrian paradise on earth[42] (Lisowska 2004).

The **Nile**, the most known example of a holy river, flows across the desert as the nourishing mother and the provider of the fertile alluvium (Margul 1995, p. 55–73). The axis of the Nile, flowing from the south to the north, along with the direction of the Sun's journey, determined the monumental layout axes not only of cites and settlements but also of the whole Egyptian world (Niemczyk 2002, p. 67). The pharaohs made offerings to the blue god of the Nile – Hapi, who ruled the regular overflows, providing the welfare of Egypt[43]. The holiness of the river survived the changes of religions and was also acknowledged by the early Christians (Margul 1995, p. 59). In addition to being worshipped as the river-nourishing mother, the Nile was considered a significant border symbol. The extent of its overflows demarcated the fertility and heath, life and death as the ribbon of the Nile was the boundary and the connection between two worlds – the underworld in the west and the upper world in the east (Niemczyk 2002, p. 67). The aforementioned nilometers – deep wells reaching the level of ground waters – not only served the measurement and cooling functions but also were "the houses of eternity", believed to be a resting place for souls leaving the grave. The illusion of the connection with the underworld was reinforced by the reflection effect during the culminant overflows of the Nile when the water reached the Abu Simbel or Karnak temples and created a mirror reflecting the architecture of the Great Hypostyle Hall (Niemczyk 2002, pp. 74–75). The holy lake in the Amon-Re temple complex in Karnak, linked with the Nile by an underground corridor, was the place where the worshipping barges arrived. The mirror of the water was accessible thanks to the stairs while the pond was surrounded by trees, invaluable in the desert environment (Niemczyk 2002).

---

42 The Persian gardens are distinctive for their *czāhar bāgh style* which refers to the Iranian cosmology. A four-part garden was divided by four canals symbolising the rivers of life. The style, imported and improved by the Muslim culture, has survived to our times and has left its strong mark on the architectural and town planning principles of the Muslim world. It is explicitly present in Moorish architecture of Iberian Peninsula, e.g. in the layout of the Court of the Lions in Alhambra.

43 Hapi had many titles i.a. "The Lord of the Fish and the Waterfowl" as well as "The Lord of the River Bringing Vegetation". The Egyptians' holy river had other gods as well: Khnum – a ram-headed master of the First Cataract, who managed the Nile overflows and protected the farmers as well as Sobek – the god with a crocodile's head – symbolising fertility – who ruled the riverine waters and swamplands.

The **Jordan,** whose waters witnessed John the Baptist baptise Jesus, also got glorified as a holy river in the Jewish-Christian culture. It is the symbol of the nourishing mother and the border river as it separated the Promised Land from the desert; while the Dead Sea, which it flows into, became the tool of God's justice when it flooded the sinful cities of Sodom and Gomorrah (Margul 1995, p. 59). The parable of Noah's Ark and the flood emphasises the punishing force of the element but, simultaneously, indicates its purifying and life-giving power as the flood causes destruction followed by exuberant revival.

The **Huang-Ho,** the holy river of the Chinese civilisation, which penetrated the edge of the Gobi desert and similarly to the Nile carried the fertile alluvium, caused disastrous floods having changed its course.

Whereas in India, the **Ganges** and the **Jamuna** have been worshipped till contemporary times. The rivers confluence places, in addition to economic and military significance, also had remarkable sacred power. Thus temples were erected and cities were founded in such places. In the Hindu city of Prayag (sacral name Triweni meaning Tri-confluence, currently Allahabad), three rivers-goddess met, namely Ganga, Jamuna and Saraswati, making the place triple sacred. The waters of the Ganges (the earthly residence of Ganga goddess and the most sacred river of India) purify sins, according to Hindu beliefs, while the burial in its current guarantees a happy rebirth or liberation from subsequent incarnations. Three holy cities Hardwar, Allahabad (Prayag) and Varanasi (Benares) attract thousands of pilgrims who perform ritual ablutions on the ghats (steps of stone slabs leading to the banks of the river). The constantly clean and bio-stimulating water is the Ganges phenomenon, considering the loads of contaminators, including ashy corpses of the dead (Margul 1995, pp. 65–66).

In Europe the confluence places were worshipped likewise. Tadeusz Margul (1995, p. 60) mentions the meeting place of the **Saone** and the **Rhone** in contemporary Lyon which was called Lugdunum (fortress of the god Lugus) by the Gaelic Celts, as well as the place where the **Main** flows into the **Rhine**, protected by god Mogons, who gave the name to the contemporary Mainz.

The European culture inherited the enormous and still alive **water symbolism** from the **Greeks** and the **Romans.** Admittedly, the Italian, Balkan and Peloponnese Peninsulas lack big rivers but the galaxy of the Greek and Roman gods, being the personification of water sanctity, is overwhelmingly rich. Its influence on the architecture and town planning is described by i.a. Ernest Niemczyk (1995, 2002) and Aleksander Krawczuk (1998) as well as many other European authors (including M. Nick 1967, D. Boeminghaus 1980, R. Tölle-Kastenbein 1990 and others).

In the Hellenic culture, every spring, stream or river had their protective gods. **Achelous**, the god of the biggest Greek river, was considered nymphs' father, and his name was used to label any fresh running water and its sanctity (Krawczuk 1998, p. 96). The classical mythology was filled with numerous nymphs who lived in seas (the Oceanids), rivers (the Nereids), springs (the Naiads) and wetland meadows (the Leimakids). The power of the seas and the oceans was symbolised by the gloomy and vindictive **Poseidon** (in the Roman culture – **Neptune**), especially worshipped in harbour cities. His mythological son, Proteus also deserves mention as one of the "Old Men of the Sea", who could change his form like water and has still been the synonym of changeability.

In addition to the worshipped real rivers, sacred mythical rivers existed as well, equally influencing the Western culture. The ocean vastness was equated to the mythic **Oceanus River** encircling the world and, according to the Greeks, determining its end. While **the Styx** represented the boundary between the world of the alive and the dead (Hades), the Styx crossing in the Charon's boat was irrevocable and swearing on the waters of the river – uncontested. In the Hades underground there flew three more rivers of remarkable qualities: the Cocytus (the icy river of moaning), the Lethe (the river of forgetfulness and oblivion)[44] and Phlegethon (the river of fire).

In ancient Rome it was popular to worship the **Tiber,** whose master **Tiberinus** was the most powerful of river gods. Pollution or interference into the river course could result in the god's fury therefore the Romans gave offerings to the tutelary deities while building bridges or reinforcing the banks[45]. The cult of springs was particularly practised, while the master of springs – Fontus personified their life-giving power. Ornate forms of the Roman fountains, architectural symbols of springs, frequently depicted the water deities protecting the sanctity and purity of drinking water, which was supplied by the aqueducts to the Eternal City (Januchta-Szostak 2017a, pp. 34–38).

---

44 The Old East Slavic religion also knew a river of forgetfulness (Zabyt' Reka), which separated the world of the living from the land of the dead. The Slavic master of the underworld, Veles, resided "behind water", while in the Ruthenian folklore – the swampland located in the central part of Nav. Source: Damian Winiarski, *Woda i demony wodne*, 25.03.2018, <https://slowianskibestiariusz.pl/zycie-slowian/inne/woda-i-demony-wodne/> [accessed: 17.06.2018].

45 Roman priests, called *pontifices*, ritually supervised the construction of bridges, protecting the peace of the rivers. The oldest Roman bridge over the Tiber was built entirely of wood without any use of metal elements in order not to harm the river (Krawczuk 1998, pp. 103–104).

The symbolism of water and river in culture and religion[46]

The awareness of complete humans' dependence on the water element, in the early stages of civilisation development, led to its sacred worship in the majority of cosmologies and religions, both polytheistic and monotheistic. It was considered as the prima materia which the universe emerged from. In primeval matriarchal cultures water personified the female element and was connected with the cult of Mother goddess, who was the life-giver, the protector of the soil and its crops, the feeder and the healer, the goddess who reigned over plants' growth processes as well as peoples' birth, life, death and resurrection. Along with the appearance of patriarchal cultures, the role of the goddess became far less significant.

In ancient Mesopotamia, before the appearance of anthropomorphic deities, the power of big rivers as well as their unpredictability and simultaneous gentleness was symbolised by bulls – the first domesticated big animals. **Enki**, who the holy ziggurat temple tower was erected for, was worshipped as a god of fresh water in Sumerian city of Eridu at the end of the 3rd millennium.

The rich **symbolism of spring** (faith, wisdom, power, hope) appears in many religions including Muslim and Christian beliefs. In Medieval Europe, the springs which had been protected by nymphs before fell under the rule of Saint Virgin Mary. Many Marian shrines were the pilgrim destinations for their "life water" healing springs, contained in marvellous architectural forms[47].

In the Muslim culture and religion, born in the desert, the water element is associated with life, welfare and paradise. It is the expression of the Absolute, pure spring of the Spirit and the Life Fountain (Sosnowski, Wójcik 2004, p. 22). The Muslim holy book – Quran – describes water as the Allah's gift, proving his almightiness and omnipresence, while the rain expresses his mercy for the mortals[48]. Water is a remarkably important symbol of the body and soul purity.

---

46 The authors of "Faces of water in culture" focus primarily on issues of the place and role of water in culture, history and literature (Burkiewicz et al. 2014).
47 Equally expressive in Christian religion was the symbol of a well – as a source of "the water of life", which, according to the Bible, is the "place of invigoration" and the source of God's blessing. "*Fons vitae* – the well of life – symbolises the holy tomb, the centre of the paradise, the Holy Spirit and, ultimately, culmination – life giving blood of Eucharistic Christ" (Niemczyk 1995, p. 104).
48 Rainmaking rituals appear in many cultures from Muslim and Hindu to the Indian community of the North America.

The Muslims perform the ritual ablution before praying, hence, pools and wells in the yards have been an indispensable element of their mosques.

The patriarchal culture of medieval Christian Europe abandoned the antique deities' pantheon, which consequently impoverished the personification of water element. However, its purifying and life giving power has been a timeless and trans-cultural attribute. In the Catholic religion, water is one of the symbols of the Holy Spirit's action. The influential power of water was mainly limited to the purification of souls thanks to the sacrament of baptism (hence the numerous and beautifully ornamented baptistries), in contrast to the body, which as a temporary and insignificant coating did not deserve a cleansing worthy of decorative architecture (although the ritual ablutions were parts of such significant ceremonies as coronations, weddings, knightly or priestly vows).

The attitude towards water was ambivalent in the Middle Ages. The purifying water power was attributed only to the holy water (hence the tradition of blessing the food and houses) whereas wild rivers and streams were considered as the dwelling places of demons[49]. The Greek and Roman deities, embodying the powers of the nature, were replaced with saint patrons. In this way the Saint John of Nepomuk became a patron of bridges, the protector from floods and drowning. According to folk tales, he protected from flood and drought, while the statues of John of Nepomuk have still been guarding roads and bridges[50].

1.7 The milestones of ancient hydraulic engineering

The history of hydraulic engineering inventions dates back to the Neolithic (ca. 5700–2800 BC). The first attempts to apply hydraulic works in agriculture had already been undertaken in ancient potamic civilisations. Naturally, they were not the proof of subjugation of the element, but were merely the first steps towards the economic or military use of rivers as well as the endeavour to fight their destructive power.

The development of the ancient hydraulic engineering began in ancient Egypt in about 3200 BC. In his *History of Hydrology*, Asit K. Biswas (1978)[51] describes

---

49 Folk beliefs expressed the fear of water by creating water monsters, while the baths in non-blessed rivers were treated rigorously. It is also possible that the first water tutelary deities from the Slavic beliefs pantheon were demonised and stayed in the folk culture as malicious vodniks, rusalkas, mavkas or samodivias.

50 John of Nepomuk's monument in Prague is well known but the figures of the saint can be found by many the bridges e.g. in Kłodzk or Lądek Zdrój.

51 In the first chapter of *History of Hydrology*, the author presents the beginnings of the hydraulic engineering of Egypt, Babylon, Palestine, Persia and China. In the

irrigation canal system dating back to that period along with the first signs indicating the level of the great waters of the Nile and so called nilometres i.e. watermarks allowing to estimate the extent of the Nile overflows, already used in 3000 BC.

One of the milestones of sanitary engineering were **water mains and sewage systems**, which started to emerge in the Bronze Age on Crete (3000–1700 BC), in the Indus valley, in the city of Mohenjo-Daro (2550 BC) and also in ancient Mesopotamia (e.g. in the cities of Babylon, Nineveh, Eshnunna and Nippur) (Bonnenberg 2009). In those cities, the fresh water-supply systems were installed along with extended underground sewage systems allowing to discharge the sewage directly into the big rivers or indirectly – using sedimentation pits, the prototypes of the first sewage treatment plants. Water mains and sewage systems were also implemented in Egypt in the mid-2nd millennium BC and in the Mycenaean civilisation approx. 2000 BC (Stanielewicz 1995, p. 157).

Aqueducts and traps improved transporting water over long distances in a topographically varied terrain. The inventions, popularised in ancient Greece (Crouch 1993)[52], not only facilitated the development of cities but also, improved by the Romans, became the symbol of controlling the water element and enriched the landscape. The construction of the **Roman aqueducts** – monumental linear structures, supported by high arcades, which were used to transport water from remote springs, was possible owing to a few other engineering inventions: the art of constructing arches, building arcade structures as well as the technology of burning bricks and producing cement. From 312 BC to 226 AD in the capital of the Roman Empire eleven aqueducts were constructed, extracting water from a few springs located even dozens of kilometres from Rome[53]. Supplying water

---

later period, especially interesting development of the hydro-meteorological science occurred in China and Korea from the 12th to the 15th century, for instance the measurement of rainfall with the use of rain gauges (Biwas 1978). It is worth mentioning that regular weather observations began in China already in ca. 1400 BC.

52 E.g. underwater Peisistratos' Aqueduct in Athens, 1,500 m long, built in 5th century BC. In the Mycenaean period, similar water mains were built in many Greek cities e.g. Tiryns in the Peloponnese and on Ithaca in the Ionian Sea (Bonenberg 2009).

53 E.g. the aqueduct called Aqua Anio Novus was supplied with water form a source 87 km east of Rome. The earliest Aqua Apia (17.6 km long, 312 BC) was underground almost at its whole length (16.8 km), while the longest one – Marcia (91.2 km long, 144–140 BC) – had only its 11.2 km long section on the surface. The underground sections of the water mains were built from special hollowed bricks called *lapides perterebrati* which were fitted together creating a hermetic channel. Roman architects precisely determined the pitch of the channels, using a device called *chorobat*. While

to the city and distributing pressurised water to different spots (e.g. fountains) required building high arcade structures, which sometimes contained several canals on different levels.

Acknowledging the sanctity of water did not contradict the attempts to understand the nature of its flow and cycle. Ancient Greece was the terrain where **hydrology** was born. According to Biswas (1978) it was initiated by: **Thales of Miletus** (6th/7th century BC) – the representative of the Ionian scientific philosophy – who perceived water as the beginning and the source, **Herodotus**, who cartographically depicted the river system and **Hippocrates** with his concept of the water element evolution. Biswas considers **Plato** (427–347 BC) the father of the ancient hydrology. He also mentions **Aristotle** (384–322 BC) who described hydrological cycle in his thesis titled *Meteorology*[54]. **Archimedes** (ca. 287–212 BC), a Greek science philosopher and mathematician, defined the hydrostatic laws in his work *On Floating Bodies*. A number of hydro-technical inventions are credited to him i.a. the Archimedes' screw, water clock and water organs, as well as the improved tackle, which was used to launch ships.

The Roman civilisation enriched hydrology not only with already mentioned aqueducts but also **Vitruvius'** concept of water cycle[55] and the first attempts to measure water flow (Biswas 1978).

The **water wheel**[56], which enabled water energy use, was the invention of key importance to the medieval Europe. **Watermills** became common in the 2nd

---

owing to the Greek invention of the inverted drain trap, they could overcome vast lower terrains. More: *Imperium Romanum – historia antycznego Rzymu* <https://www.imperiumromanum.edu.pl/kultura/architektura-rzymska/budowle-rzymskie/akwedukty-rzymskie/> [accessed: 16.07.2018].

54 Aristotle discovered the phenomenon of water vapour condensation. He noticed the connection between the amount of water in streams with evaporation and insolation.
55 Vitruvius (Marcus Vitruvius Pollio) – a Roman architect, civil and military engineer, living in the 1st century BC, devotes the whole 8th book to water in his treatise *The Ten Books on Architecture* (Vitruvius 1999). He not only discusses the subject of the water cycle in the nature but also the methods of its extraction and transport to the settlements and cities. In the 15th century Leonardo da Vinci accepted Vitruvius' theory and developed the first, commonly accepted scheme of hydrological cycle in his notes.
56 Strabo, a Greek historian, geographer and traveller, first mentioned a working watermill at the beg. of the 1st century BC, in the region of Kabira in Pontus by the Black Sea, in the outskirts of the Roman Empire (Orłowski 1993, p. 48). Also Vitruvius confirms the existence of mills powered by water wheel in the 1st century BC. But the 3rd century projects by a Greek mechanic Philo of Byzantium prove that the water wheel had been already known much earlier. More: Ekologia. pl <https://www.ekologia.

and 3rd centuries AD. They were built on fast flowing rivers and streams in the form of cascade rows in order to maximally use the force of water fall[57]. Water wheels were applied not only to grind the grain and extract oil, since 12th century but also used in forges, sawmills, fulleries, spinning mills and other craft and industry workshops. Water energy became a catalyst for economic development and urbanisation while the watermills were gradually replaced with steam machines only in the 19th century during the Industrial Revolution, and later – with combustion and electrical engines.

After the collapse of the ancient civilisations, water supply and sewage systems of the ancient cities passed into oblivion although the remaining sections of the Roman aqueducts still worked locally during the Renaissance. In the Middle Ages, water was extracted from rivers and streams or city wells. Even during a siege, minor cities, surrounded by walls, could use water from the wells which reached the water-bearing layers as well as rainwater containers[58]. The first wooden **water conduits** began to appear for the needs of burghs[59]. Water was drawn with the use of chain buckets, powered by water wheels, and lifted to the height of several or several dozen metres to the top of the water tower to be subsequently distributed by pipes around the city. The faeces were retained in cesspools while the wastewater flowed in open gutters, lining the streets, directly to natural reservoirs – moats, rivers, streams and lakes. Torrential rains helped clean the streets as precipitation water flowed in the same canals. Only in some cities of Arabian or Roman origin, underground sewage systems were preserved. The Roman Cloaca Maxima[60] was still used in contemporary times. The

---

pl/ciekawostki/mlyny-wodne-antyczna-technika-w-swojski-klimat-wpisana,18210. html> [accessed: 16.06.2018].

57 E.g. the commonly known Roman watermill complex in Barbegal near Arles in France, consisting of 16 water wheels.
58 Original rainwater retention solutions were used in Venice (more: Kaminski 2000, p. 224). Wacław Ostrowski devoted a chapter to the analysis of Venetian urban ecosystem in his book *Wprowadzenie do historii budowy miast*, 2001, pp. 26–34.
59 In some Polish cities such installations were constructed already since the 13th century, e.g. in Opole and Racibórz (1258), in Wrocław (1272), Poznań (1282) and in Mydlniki near Kraków (1286) [Żukow–Karczewski, n.d.]
60 Cloaca Maxima – the main sewage canal in Rome – whose construction was initiated by Tarquinius Priscus in the 7th century BC in order to dry the wetlands in the valley between the Esquiline, Viminal and Quirinal Hills. Already existing stream course was deepened and its banks were reinforced. The waters flowed to the Tiber. Initially, the canal was open but in 184 BC Cato the Elder had modernisation works performed and the canal was closed. In the following years, the canal underwent numerous

biological contaminants had not caused ecological disasters until they exceeded the level of the rivers self-purification capability.

The technology of digging canals, building harbours and boats enabled the development of navigation already in antiquity. But the invention of the **lock chamber**[61] became a milestone in the process of transforming rivers and constructing inland water routes in Europe since it enabled vessels to overcome the differences in water level. The canal lock chamber was invented in China in 984 by Qiao Weiyue and built for the first time on the Western river near Huaiyin (Temple 1994, pp. 196–197). The first known lock chamber in Europe was built in Holland in 1373[62] and, rapidly popularised, opened the way to the development of inland waterways system.

---

    renovations, redevelopments and sometimes changes of its course. The canal is ca. 600 m long, its headroom is 4.20 m, and width 3.20 m. It is the oldest contemporarily used canal in Rome (Krawczuk et al 2005, p. 215). Source: <https://pl.wikipedia.org/wiki/Cloaca_Maxima> [accessed: 16.06.2018].

61 Earlier, the Chinese used the decline which flat-bottom boats were pulled on, with the use of ropes, strained on hoisting winches, turned by oxen. The first 76.2 m long lock chamber with two gates was covered with a roof and made it possible to overcome the level difference from ca. 1.22 m to ca. 1.50 m. The invention was rapidly popularised in China facilitating water transport development and the construction of i.a. the Grand Canal and the redevelopment of the Lingqu ("Magic Canal") in 11th century (Temple 1994, pp. 196–197).

62 In 1373, a lock with a big chamber, used to raise and lower a few boats, was built in Vreeswijk on the canal connecting Utrecht with the Lek River. Similar locks were constructed in Delfshaven, Brill and other places from 1390 till 1430. Already in 1391, there were two locks in Germany, on the waterway connecting Hamburg with Lübeck. The first description of a lock in a technical book was provided by Leon B. Alberti in 1485. In England, lock chambers appeared in ca. 1550, in Sweden only in ca. 1600. source: Wynalazki i odkrycia < http://kanalgliwicki.net/sluzy/> [accessed: 15.07.2018]

2 The CONQUEST period

> *The "control of nature" is a phrase conceived in arrogance, born of the Neanderthal age of biology and philosophy, when it was supposed that nature exists for the convenience of man. The concepts and practices of applied entomology for the most part date from that Stone Age of science.*
>
> [Rachel Carson 1962]

The conquest period was gradually developing along with losing respect for the power of water element. In the Middle Ages water flowed from the pedestal of an idol to the role of valuable but obstructive matter. Popularisation of the Christian religion led to disappearance of "pagan" beliefs in the rivers sanctity. The conviction, based on the Biblical words, that the man (male!) is "the king of all creation", while the whole earth with all its resources was given to him to be subdued[63], justified predatory exploitation of the resources and absolute subordination of the nature to human needs.

At the turn of the 15th century, Europeans faced "big water" barrier when entering the era of oceanic civilisation and colonisation; while in the periods of Renaissance and Baroque "small water" became a material for constructing vast fortification complexes and sophisticated garden arrangements. Broadening the hydrological[64], hydrographical and nautical knowledge, as well as the development of boatbuilding, shipping and hydraulic works, made water an enemy which could be dominated. Technical progress gradually reduced previous

---

63 The divine right of kings – a philosophical-political concept spread in Western Europe in the 17th and 18th century, justifying absolute monarchs' power by its divine origin, initiated by *Patriarcha* by Robert Filmer (1680). It also referred to the power over the nature. According to Filmer, the Creator gave Adam the right to rule the whole creation, thus the subjugation process of the wild nature was approved by God.
64 Potamology, (from Greek ποταμός potamós = "river") – geographical science examining rivers, is one of the disciplines of hydrology (more: Bajkiewicz-Grabowska, Mikulski 1999, p. 14).

limitations concerning the use of water sources[65] and raised the confidence that controlling the water element is only a question of time and expenses.

2.1 Challenging water – the Dutch phenomenon

The Dutch created a specific culture of coexistence with water: they were precursors of its conquest but simultaneously held it with respect. The beginning of the settlement on the terrain of contemporary Netherlands dates back to the 4th century BC (Balicki, Bogucka 1989). Fertile but wetland areas by the estuaries of the Scheldt, the Meuse and the Rhine, threatened by the sea and river storms, posed an immense challenge to the settlement.

For that reason, Holland[66] became a European pioneer in the field of reclamation and hydraulic engineering. The medieval inhabitants of the Netherlands used to say: "Deus mare, Batavus litora fecit" which means: "God created the sea, but the Dutch created the shore" (Bogucka, 2011). From the 1st till the 11th century so called terps were piled up, which were artificial sandy mounds, up to 10 m high, where initially single farmhouses were located, and later – the whole cities. Also, the seaside sand-spits were heightened and reinforced, followed by subsequent earthen dykes construction. "They were first low embankments of single farmhouses, later villages and individual settlements, and, ultimately, connecting them together created a sequence of embankments, separating the entire riverside or seaside lowlands from the invasion of high swollen waters" (Makowski 1997, p. 199).

---

65 David Blackbourn, a British historian, in his book *The Conquest of Nature* (2006) illustrated the tools and the degree of interference in the nature in the early modern period on the example of Germany. As Stanley Hoffman, the reviewer, says: "This conquest of nature was seen as a peaceful victory of science and technology; in reality it was often "the handmaiden of war". The "water wars" the book describes, even when they were not at the service of military designs, created their own conflicts: they "set rival users against each other" and local interests against larger ones, who usually prevailed". *Source:* <https://www.foreignaffairs.com/reviews/capsule-review/2006-09-01/conquest-nature-water-landscape-and-making-modern-germany> [accessed: 1.08.2018].

66 Contemporary Netherlands came into existence in the 17th century, as the Northern Netherlands, on a tiny scrap of land of the depression coast of the North Sea. Already in the 16th century the country was very densely populated; the country of free peasants and resourceful burghers, where nearly half of the population lived in cities (Bogucka 2011, p. 61)

Stages of polders reclamation

Creating the inhabitation space required the construction of reclamation systems, sequences of embankments, canals and locks, which in the period from 1100 to 1700 changed the Dutch landscape into a hydraulic system (Hooimeijer et al 2005). The Dutch began extending and reinforcing their territory by draining the swamplands, building dykes and settlements.

The invention of the windmill, used in the 16th century to pump water out, made it possible to regain bigger areas, while the engineers: Jan Adriaensz Leeghwater, Cornelis Cornelissen and Simon Stevin became the creators of the Dutch landscape. At the beginning of the 17th century, the Beemster Polder was created to the north of Amsterdam by Jan Adriaensz Leeghwater (1575–1650), windmills constructor. An area of 72 km^2, located 3.5 m below the sea level was surrounded with 45 m long embankment and draining it with 50 windmills took five years (Hooimeijer et al 2005).

At the end of the 16th and the 17th centuries Dutch technology enjoyed primacy in water and military engineering. The city plans of Naarden (1579), Klundert (1583), Willemstad (1583) or Breda (1635) were based on the principles of fortification developed by Simon Stevin, and between 1647 and 1704 Menno van Coëhoorn published *Nieuwevestingbow*, laying foundation for New Dutch School of fortification. The plan of an ideal harbour city by S. Stevin (1590, Fig. 2.7), based on the network of perpendicular streets, surrounded by defensive bulwarks and moats, was applied not only to construct cities but also to newly created polders like Beemster (1612), Schermer (1635) or Watergraafsmeer (1629), which were divided into rectangular plots by canals and roads.

In the 17th and the 18th centuries the Dutch efforts concentrated on gaining new areas for building landed estates, fortresses, canals and water routes. Already in the 18th century the density of the reclaimed rivers and artificial shipping canal system in the Netherlands exceeded 10 km per 100 km^2 (in contrast, at present – 20 km/100 km^2, Wojciechowski 2000, p. 191), whereas the technological progress gradually enabled the shift in transformation trends: from the defence against water to active transformation of water structures.

The subsequent stage of gaining land included the big lakes. The Haarlemmermeer Polder (Fig. 2.8), 180 km^2 of surface area, was drained in the years from 1848 to 1852 with the use of steam pumping. In the 19th century new terrains were gained mainly for the needs of intensively developing industry, railways and motorisation (Hooimeijer et al 2005).

The conquest also involved sea waters (Fig. 2.9). The first plans to close and drain the Zuiderzee terrain date back to 1667. However, level of technology at

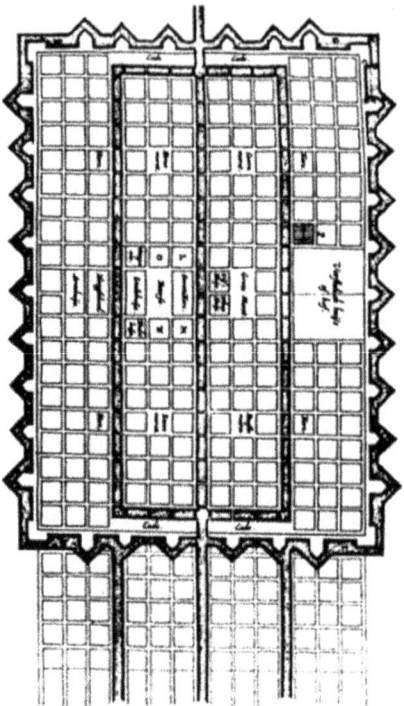

Fig. 2.7: The plan of an ideal harbour city by Simon Stevin (1590), source: Kostof 1991.

the time was not advanced enough to undertake the venture. Only after the steam pump had been invented at the end of the 19th century, did the engineer Cornelis Lely suggest building a dam closing the bay and in 1891 he submitted a development plan covering four big polders. The big dam of Afsluitdijk[67], built between 1927 and 1932, separated the Zuiderzee Bay from the North Sea, changing it into an inland lake called IJsselmeer. The dam allowed to gradually drain the former bay, within which by the end of the 1960s three polders had been created, namely: Noordoost Polder (in 1942 surface area of 48,000 ha), Oostelijk Flevoland (in 1957 – 54,000 ha) and Zuidelijk Flevoland (in 1968 – 44,000 ha).

---

67 At the end of the 19th century, an engineer Cornelis Lely developed a project of a huge, 32 km long Afsluitdijk dam (literary: Enclosure Dam). It stretches from Den Oever in the North Holland province to Zurich in Friesland. It is 90 m wide and 7.25 m high.

Challenging water – the Dutch phenomenon 51

Fig. 2.8: Dutch topographical map of a polder Haarlemmermeer (2015), source: Janwillemvanaalst, <https://en.wikipedia.org/wiki/Haarlemmermeer> [accessed: 13.01.2019].

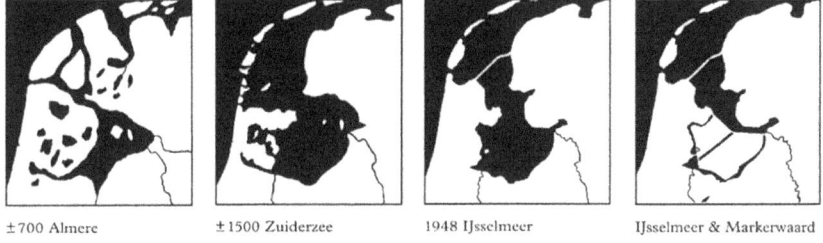

±700 Almere   ±1500 Zuiderzee   1948 IJsselmeer   IJsselmeer & Markerwaard

Fig. 2.9: The transformation stages of Zuiderzee Bay, source: Hooimeijer et al. 2005.

As a result, the surface area of the small country was extended by 142,000 ha. Currently, as much as two-fifths of the surface area of the Netherlands consists of polders. C. Lely also planned the fourth polder of Markerwaard, however, despite quite an advanced stage of construction, the project was forsaken in the 21st century.

Amsterdam, as a water city phenomenon (Fig. 2.10), has inspired numerous scientific publications. It is still worth emphasising though how closely its urban structure relates to water. The name of the city derives from a dam on the Amstel river (*Aemstelredamme*). Its specific, semi-circular layout, determined by multiple canals was created in the 17th century – "Golden Age" of Amsterdam

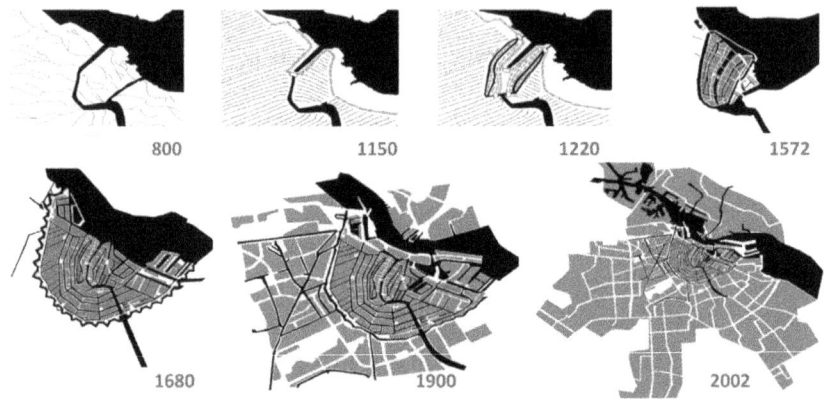

Fig. 2.10: The stages of Amsterdam development. Developed on the basis of historical maps and the book: Hooimeijer et al. 2005.

which was developing owing to the profits from the Dutch colonies and international trade. The subsequent stages of development followed: Plan-Kalff (1876), Plan Zuid (project by Berlage, 1920–1940) and General Extension Plan (1935), Bijlmermeer plan (1970) and the plans of colonisation and redevelopment of the coast (since the 1990s).

Olęder settlement in Poland

The Dutch skills in hydraulic engineering were used in the entire Europe. Olęders, originally the settlers form Friesland and the Netherlands (in the 16th and 17th centuries), significantly contributed to the processes of draining and developing wetlands as well as transforming the hydrological system in northern Poland. The first development attempts were made in Żuławy Gdańskie, located in depression areas of fertile but often flooded land[68]. The Olęder settlers who knew reclamation technology and very highly developed agriculture were an extremely desirable group of colonisers thanks to which they founded the

---

68 The embankments of Żuławy Wiślane and Elbląg, constructed between 1288 and 1294, are attributed to Meinhard von Querfurt, a Landmaster (Landmeister) of Teutonic Order (Makowski 1997, p. 201).

villages in accordance with special rights[69] in Royal Prussia, along the Vistula river and its tributaries, in Kuyavia, Masovia and in Greater Poland[70].

It is worth mentioning that **the process of Olęder colonisation of wetlands was performed with great respect for the water element**. The settlers did not separate themselves from the rivers but rather adapted their houses and farmhouses to the changeable levels and conditions of water flow. They built houses on terps, artificial sandy mounds, and elevated roads, enabling the inhabitants to reach their settlements during water level rise. The buildings in the floodplains were situated along longitudinal axis, parallel to the watercourses, which provided protection from overflow force. While between the settlement and the river – they shaped the terrain in such a way that facilitated the water flow and runoff (low embankments, canals and dykes). They used transporting functions of watercourses constructing bidungas[71] on the banks of rivers and canals. Olęder settlements featured fascine fences, willow cultivation as well as rows of willow-poplar vegetation lining the roads, canals and balks (Zaraś-Januszkiewicz et al 2013).

The idiosyncratic architecture of the Olęder settlements resulted from the adjustment to the reality of **living with flood**. The living and the livestock parts were usually connected under one common roof. Such a solution took less space and enabled the farm work almost without the need to leave the constructed shelter. The main living space was located upstairs which protected the inhabitants from being surprised by flood during sleep. Additionally, wide stairs enabled them to bring the livestock to the attic in an emergency (Szałygin n.d.). "The living part of the farmhouse was directed towards the river. In front of the living part, small decorative-usable gardens were located. They served both decorative and fruit-growing purposes but they also retained the fertile flood outwash. Flood waters pouring into the farm were used in two ways. They were supposed to fertilise the fields equipped with the rows of trees and bushes as well as fascine fences installed between them. The second benefit was, in a sense, cleaning. The Olęders welcomed flooding waters with an open door, as it were. The water washed the sewage from the livestock part and carried

---

69 The settlers were entitled to the privileges on the basis of the law used by the Frisian and Netherlandish colonisers (e.g. personal freedom, multiannual or perpetual usufruct, possibility to pass the land on to the heirs) (Baranowski 1915).
70 More: Baranowski 1915; Rusiński 1947.
71 Binduga is a place used to prepare wood to float and also a river berth.

the fertile outwash to the fields. Therefore the living parts were located in the front so that the water could flow into the farming part afterwards." (Zaraś-Januszkiewicz et al 2013).

The Dutch influence is also explicitly apparent in the Polish cities which flourished owing to the thriving trade between Poland and the Netherlands in the 16th and the 17th centuries. Royal Prussia, especially Gdańsk along with Wisłoujście and Elbląg, received embankments from vast earthen banks connected with broad moats filled with water, specific to so called Dutch old school of bastion fortifications which became popular in the first half of the 17th century (Bogucka 2011, p. 70).

The layout and the architecture of Gdańsk in that period were so similar to Netherlandish town planning that it was called the other Amsterdam. Not only the close functional-spatial relations with the Motława, the port berth, fortifications and canals but also the houses of small red brick and public edifices (the Main Town Hall, Artus' Court, Armoury, Golden and Green Gates, Old Town Hall as well as the churches) resembled Amsterdam[72]. Through Gdańsk, the Netherlandish influences reached Toruń[73] and further the central and southern Poland: Masovia (Warsaw), Lesser Poland (Cracow, Zamość, Lwów, Przemyśl, Rzeszów), and even Silesia (especially Wrocław, Nysa, Brzeg, Chrzanowski 1995).

Olęder settlers enriched Poland with their centuries-old experiences of coexistence with water and highly developed hydraulic engineering culture, which featured respect for the element and its rights, the knowledge of the laws of the flow but also implementation of hydraulic engineering inventions and collective effort taken in order to maintain the water systems. The Dutch have been aware that challenging water demands constant effort, indispensable to maintain the "conquered" terrains. Otherwise, the element will immediately claim its space.

---

72 The authorities of Gdańsk and big cities of Royal Prussia employed the Netherlandish specialists as the city engineers, conservators of canals and embankments. More: Bogucka 2011.

73 At the turn of the 16th century, the prosperity period of the city, the Main Town Hall was redeveloped in accordance with the design by Anthonis van Obbergen while the facades of many tenements and corn houses were changed. Similar works were performed in Elbląg as well (Bogucka 2011).

2.2 Navigation and industry – catalysts for urbanisation

Transoceanic ventures – the colonial period

First expeditions across the Atlantic Ocean at the turn of the 1st millennium AD were undertaken by the Vikings who, using a primitive version of a compass[74], reached Iceland, Labrador and Newfoundland where they developed small settlements. However, the real civilisation breakthrough was initiated by the major expeditions undertaken by Portuguese and Spanish sailors at the turn of the 15th century. The great geographical "discoveries" by: Prince Henry the Navigator, (1419–62), Christopher Columbus (1492–1504), Vasco da Gama (1497–98) and Ferdinand Magellan (1519–22), opened the way for the expansion of European culture and changed the Europeans' mental image of the world.

Consequently, the Renaissance Europe entered the oceanic civilisation period (Ocioszyński 1968) and started the "cross and sword" conquest, creating the foundation for colonialism[75]. In addition to the political domination, colonialism inextricably entailed the phenomena of expansion, exploitation, control, and competition, which showed not only in the attitude towards the overseas non-Christian cultures[76], but also in the way the natural resources and environment were ill-treated.

---

74 The compass was a Chinese invention. The first description of a compass, nearly in contemporary form, was provided by a Chinese scientist, Szen Kua. While the mention of using the compass by sailors on the Mediterranean Sea dates back to 1190. About 1269, Petrus Peregrinus, a French explorer and scientist, implemented 360 degree compass face. It was popularised in Europe at the end of the 13th century and became one of the main technological advancements (not counting hinge rudder) which enabled the ocean-going navigation and led to the age of big geographical discoveries. Source: Wynalazki i odkrycia < http://wynalazki.andrej.edu.pl/index.php/wynalazki/24-k/284-kompas> [accessed: 15.07.2018].

75 Colonialism – a historical phenomenon which consists in subjugating and maintaining the political and economic control by one country over another with the aim of their exploitation. *Encyklopedia PWN*, <https://encyklopedia.pwn.pl/haslo/kolonializm;3924014.html> [accessed: 15.07.2018]

76 The Spanish expansion, following the discovery of America in 1492, caused the destruction of the developed Indian civilisations: Aztecs (in the area of the contemporary Mexico and Guatemala), Incas (in the Andes, in contemporary Peru and Chile) op. cit.

The colonisation of the coasts of Africa, Central and South America as well as part of North America by Spain and Portugal, became the source of their political power and material riches. In 1494, Spain and Portugal agreed on the division of the beyond-European world between the two of them but their colonial monopoly was questioned by other powers: France, Netherlands and England, which began to compete for colonies at the turn of the 16th century. In 1524, French expeditions reached the eastern coast of North America and since the second half of the 16th century the sailors from England and Netherlands appeared on the sea routes. The Dutch colonised south-eastern Asia and Oceania, while the English attempted to find sea routes to the countries of the East along the northern coasts of Eurasia and America. The sea coasts and the rivers had strategic significance. The seaside settlements were colonial beachheads while the rivers were mainly used to penetrate the subjugated land and transport the local riches and slaves.

In Europe, the extension of the sea harbours and docks, intensive trade and cultural exchange as well as the inhabitants' wealth increase contributed to flourishing of the Italian, Spanish, Portuguese, English, French and Dutch harbour cities. The most important trade cities in the first half of the 16th century: Genoa, Lisbon, Antwerp, Lyon and Augusta considerably raised their material status and enriched their architectural image in that period.

The water routes and related transport and trade, also facilitated the advancement of the cities located by bigger rivers. **In the Netherlands** – Amsterdam, a significant sea harbour, developed by the Amstel river, Rotterdam by the Meuse estuary, Utrecht by the Rhine, Haarlem by the Spaarne river. "At that time, the Netherlands were the terrain of the crucial river and road junctions. Ships sailed on the Rhine, the Meuse and the Scheldt to numerous sea harbours and back – up the river" (Stanielewicz 1995, p. 163). In the mid-17th century, the Netherlands assumed the leadership in the world's economy owing to the development of sea and oceanic shipping, while Rotterdam developed from a small fishing settlement into a buoyant harbour city and became the most important sea port between Antwerp and Amsterdam. Its specific triangular layout with wide Boompjes waterfront resulted from city adaptation to intensive water transport and trade in goods (cf. Chap. 4.2).

In the 17th century England also became a sea power, while water transport – the source of **London** inhabitants' wealth (cf. Chap. 4.3). Loading berths, shipyards and stores developed along the Thames, which changed the seafront into a long harbour zone. For the Londoners, the Thames was not so much the source of beauty as wealth (Ostrowski 2001). Whereas France became the leader of canals construction[77]. Unlike the London Thames, the Seine in Paris turned into a representative water route and the city showcase.

The prosperity of the sea trade also resulted from the increase in the inland transport. During the Renaissance Poland was a country where riparian transport was used on a large scale. Throughout the whole 16th and half of the 17th centuries, it was the main supplier of crops and forest produce which were floated to **Gdańsk** on the **Vistula** river. Numerous cities developed along the Vistula route, namely Połaniec, Sandomierz, Kazimierz Dolny, Stężyna, Warszawa, Zakroczyn, Płock and Włocławek. "In the discussed period, the Vistula became a transport artery with the highest intensity of freight in the world. Shipping on the Rhine achieved similar extent only in the 19th century" (Stanielewicz 1995, p. 164).

The Industrial Revolution and development of waterway systems

The process of intensive technological, economic, social and cultural changes, initiated in the 18th century in England, which followed the transition from the agriculture- and craft-based economy to the mechanical factory production on the industrial scale, was labelled as the Industrial Revolution[78] and became a breakthrough stage in the European history. The catalysts for the watershed were the advancements in shipping and science, whose achievements were implemented in different branches of economy.

The period of the 18th and the 19th centuries in Europe can be easily called the **century of water engineering**. Admittedly, in England, France, Germany, the Netherlands and Northern Italy, the inland water transport had been developing since the Middle Ages, however, the majority of the buildings connected with canals and navigation were constructed in the period between 1700 and 1835 (Wojciechowski 2000, p. 191). Redeveloped locks and boat lifts[79], harbour

---

77 In 1684, The Canal du Midi (241 km) was built, connecting the Bay of Biscay with the Mediterranean Sea, which shortened the route from the Mediterranean harbours to the western European countries and contributed to the development of the Bordeaux harbour.

78 The term "Industrial Revolution" was first used in 1884 in the title of the book by Arnold Toynbee, who features the period of 1760–1830 as the darkest period in the history of England. More: Industrial Revolution – Britannica Online Encyclopaedia <https://www.britannica.com/event/Industrial-Revolution> [accessed: 15.07.2018].

79 A boat lift vertically lifts a bath with water and a ship from low to high level or the other way round. James Anderson pioneered the first boat lifts, publishing his project in Edinburgh in 1794. Seven functioning boat lifts were constructed from 1834 to 1867 on the Grand Western Canal, also in England. They lifted vessels of 8 ton carrying capacity. In 1875 Anderton boat lift was constructed near Northwich. Source: *Od przewłoki do*

devices, structures and ship propulsion, reinforced the "belief in advancement" and confidence that it is possible to subjugate the element with the means of water engineering. Furthermore, a number of inventions, including the landmark invention of the steam engine by James Watt[80], streamlined the production of spinning mills and looms, which initiated the development of textile industry[81]. Consequently, the demand for cheap water transport rose dramatically and so did riparian location of the industrial plants.

The energy crisis, deepening in England since the end of the 16th century, made the British intensify the bituminous coal mining. Its easily accessible deposits were located in the vicinity of rivers which enabled cheap and mass transport of the raw material (Freese 2016, pp. 31–32). Already in the second half of the 18th century, a multilevel, 156 km long ring **canal system** in **Cheshire** (1759–1795) was constructed. Provided with 92 locks[82], it facilitated coal consignments and goods transport. In 1776, the road between **Manchester** and **Liverpool** was opened and in 1795 the connection with Leeds and the Liverpool Canal complemented the system. The canal enabled the textile industry advancement since it functioned as the main floating route for the produced goods as well as raw materials till 1974. Along with the construction of the canal, numerous marinas (Castlefield in 1765), factories, warehouses, silos as well as housing estates for the working class appeared[83].

The 58 km long **Manchester Canal**, built one hundred years later (1887–1894), owing to its connection with the Liverpool Bay and streamlining the communication with the British colonies, became "the window to the world" for Manchester, a city remote from the coast. Liverpool and Manchester harbours, connected by the Manchester Canal, became the biggest group of ports, just after London.

---

podnośni. Jak na przestrzeni wieków ewoluowały urządzenia i budowle hydrotechniczne ułatwiające żeglugę i umożliwiające tworzenie sieci wodnych dróg śródlądowych. <http://kanalgliwicki.net/sluzy/index.html> [15.07.2018].

80 James Watt is considered the inventor of the steam machine since he developed the atmospheric engine in 1763, earlier constructed by Thomas Newcomen.

81 Since 1771, a network of spinning mills commissioned by Richard Arkwright, an English industrialist, was situated along the rivers and initially powered by water mills, and later – steam engines.

82 More: The Cheshire Ring: Northwest Canals Inland Waterways around Greater Manchester and Cheshire <https://www.manchester2002-uk.com/maps/canalss-map/> [accessed: 15.07.2018].

83 Op. cit.

The construction of the canals and river engineering determined not only the functional – spatial structure of Manchester but also of many other industrial cities in the entire Europe. Admittedly, the fast development of railway connections in the second part of the 19th century considerably diminished the significance of inland navigation but the process of radical transformations of river valleys, which involved the entire Western Europe, had caused the changes in the landscape and priority functions of rivers in rapidly developing cities. At that time, the environmental consequences of the transformations were not diagnosed and irrelevant while taking investment decisions.

The 19th century also brought a number of electrical inventions and, as a consequence, rising demand for water energy use[84]. The first hydropower station was constructed on the Fox river (the USA) in 1882, and as early as at the beginning of the 20th century, hydropower stations provided most of electrical energy in the United States (Palmer T. 1986).

The subsequent stages of the Industrial Revolution (the second one – at the turn of the 19th century and the third one – after the World War II) featured increasingly bigger contribution of science and implementation of its achievements on the industrial scale, which resulted in unprecedented increase in the use of the natural resources and fossil fuel.

2.3 River engineering

The three biggest river basin areas within the European Union territory belong to the Danube (817,000 km^2), the Rhine (185,000 km^2) and the Vistula (194,000 km^2). Their combined surface area covers over a quarter of the EU surface area (27 %); however, only the Vistula has retained its natural character within the section of over 300 km[85]. The Rhine is a regulated and busy water route,

---

84 In 1800, Alessandro Volta, an Italian pioneer in electricity, constructed a galvanic cell (voltaic cell), and at the end of the 19th century electrical current was already widely used. Nevertheless, the popularisation of electricity on a mass scale followed the invention by Thomas Edison, who patented the light bulb in 1879 and was the founder of the first public power plant, built in New York between 1881 and 1882. The power plant had six direct current generators, all of which were powered by a steam engine. More: *Wynalazki i odkrycia* <http://wynalazki.andrej.edu.pl/index.php/wynalazki/30-p/474-prad-elektryczny> [accessed: 16.07.2018].
85 The middle section of the river course form Sandomierz to Płock is currently a protected area within the framework of Natura 2000.

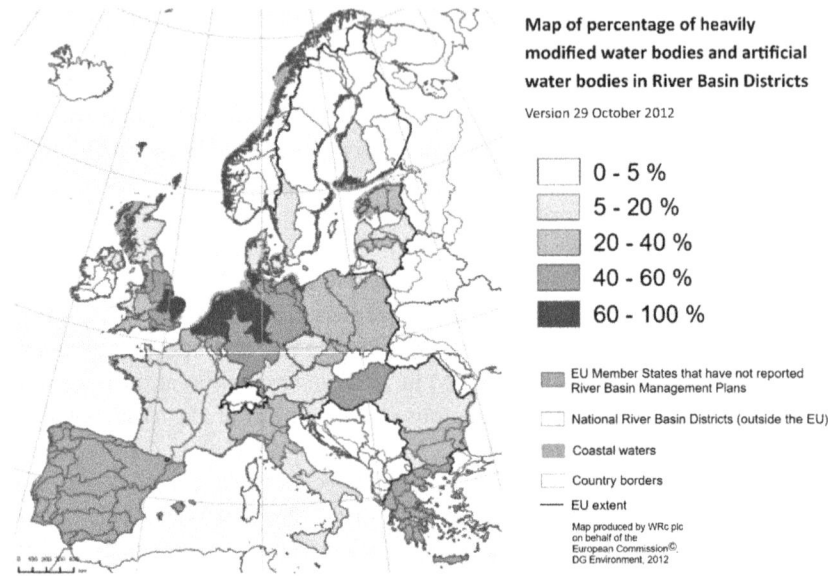

Fig. 2.11: A map illustrating the percentage of heavily modified and artificial water bodies in river basin districts, source: Map of percentage of heavily modified water bodies and artificial water bodies in River Basin Districts Version 29 October 2012 <http://ec.europa.eu/environment/water/water-framework/facts_figures/pdf/Heavily%20modified%20water%20bodies-2012.pdf> [accessed: 30.07.2018].

canal – connected with navigational rivers: the Saône and the Rhône, the Marne and the Seine as well as the Scheldt, the Weser, the Elbe and the Ems[86].

The Rhine waterway

The Rhine basin covers the most transformed water terrains (Fig. 2.11) and the most densely populated and urbanised areas of Europe. The water-economic demand of Germany required the Rhine engineering, which was performed in the middle river course between 1817 and 1876. The subsequent stage

---

86 Regular navigation for ships of up to 2,000 ton carrying capacity is possible on the length of 868 km. After the Rhine – the Main – the Danube canal construction along with the Grand Canal de Alsace, the total length of the waterways in the Rhine Basin reached 3,000 km. Source: Encyklopedia PWN <https://encyklopedia.pwn.pl/haslo/Ren;3967052.html> [accessed: 30.07.2018].

(1928–1977) involved regulation of the upper course of the Rhine from Basel to Iffezheim as well as constructing ten weirs.

The improvement of the shipping conditions along with the flooding "safety" enabled intensive economic development. **The Ruhr district**, the biggest industrial zone in Europe, was situated in North Rhine-Westphalia by the Ruhr river, a tributary of the Rhine[87]. Heavy industry developed there owing to the bituminous coal and zinc ore deposits[88] while the dense system of land routes (railway and car) as well as waterways (the canals: Datteln-Hamm, Dortmund-Ems, Rhein-Herne, Wesel-Datteln) made it possible to communicate with the North Sea as well as the Netherlands, Belgium and France. The inland water routes also facilitated the shipment of petroleum from the harbours of Wilhelmshaven and Rotterdam to oil refineries and petrochemical plants in Ruhr district, also one of the largest inland harbours in the world was built in Duisburg (transhipment 21 m t, 1999). Numerous cities[89], linked by transport systems, created the biggest German conurbation whose population is estimated at ca. 10–12 m inhabitants.

The **Rhine engineering process** required plentiful digs aiming to separate the meanders which inhibited the navigation of big ships. The river banks were reinforced and embanked along almost their entire length, radically changing

---

87 The official surface area of the Ruhr district amounts to 4,435 km^2, population: 5,172,745 inhabitants (on 31.12.2009), population density: 1,166.3 person/km^2. Source: <https://pl.wikipedia.org/wiki/Zag%C5%82%C4%99bie_Ruhry> [accessed: 30.07.2018]. Huge industrial chemical, paper and steel plants were located by the Rhine which had caused very strong pollution of the river water by the 1970s. Source: *Encyklopedia PWN* <https://encyklopedia.pwn.pl/haslo/Ren;3967052.html> [accessed: 30.07.2018].

88 The biggest mining centres are cumulated along the canals: The Rhine-Herne Herne (Duisburg, Gelsenkirchen, Recklinghausen) and Dortmund–Ems (Dortmund). Thermal power stations powered by burning coal are the basis of the energy technology; the most highly developed branches of the processing industry are: ferrous metallurgy (Duisburg, Dortmund), machine building and metal industry, especially the production of the machines and devices used in mining and metallurgy, as well as machine tools, metal structures, heavy cranes and pipes (Essen, Duisburg, Dortmund), car industry (Opel w Bochum), electrical engineering and electronic industry and precise mechanics (Essen, Dortmund); aggregation of oil refinery industry, petrochemical, chemical coke and carbo-chemical industrial plants. (Gelsenkirchen, Dinslaken, Marl). Source: *Encyklopedia* PWN <https://encyklopedia.pwn.pl/haslo/Ruhry-Zaglebie;3969965.html> [1.08.2018].

89 The cities of Dortmund, Essen, Duisburg, Bochum, Gelsenkirchen, Oberhausen, Hagen, Hamm, Mülheim an der Ruhr, Herne, Recklinghausen, Bottrop along with the neighbouring cities of: Wuppertal, Düsseldorf, Mönchengladbach, Leverkusen, Köln and Bonn.

the morphology of rivers and flow conditions. The effect of the engineering was "straightening" the river and shortening its course by 81 km. The construction of the embankments narrowed the valley and reduced the flood plain area (from 1000 km^2 to 140 km^2), which increased hydric erosion as well as raised and accelerated the flow of the flood waves[90].

The research conducted in the Rhine Basin (Schultz 1995, p. 216; Klaiber 1996, p. 396–400) showed that as a result of the decreasing valley retention[91] and hydraulic engineering development[92] of the upper Rhine, flood whose estimated frequency was once every 100 years at the beginning of the 20th century occurred six times during the past century; while the cities threatened with flood occurrence once 200 years before 1955, now are at flood risk even every 50 years (Wawręty and Żelaziński 2007, p. 23).

The impact of urbanization on the increase in flood hazards can be well illustrated by the example of the Ems Basin (a tributary of the Rhine in the Ruhr district) where twice as much ground has been sealed for 50 years than in the whole previous settlement history of this area. Consequently, six floods with very high pick flow occurred in the 1980s (cited in Geiger and Dreiseitl 1999, Londong 1993, 1994).

The cumulated runoffs from highly urbanised basins caused high overflows and floods on the Rhine[93]. In Germany, the most affected cities were Koblenz, Bonn and Köln, and in the Netherlands ca. 500 km^2 were covered with water during the flood in 1993. Repeatedly occurring extreme flood phenomena in

---

90 The upper course of the Rhine (below Basel) was shortened from 354 km to 273 km. In the section from Basel to Karslruhe, the flood wave flow time decreased from 64 h to 23 h, which increased the danger of cumulating the waves from the tributaries and the Rhine. Such a phenomenon occurred in January 1995. Source: Janusz Żelaziński, n.d., *Techniczne środki ochrony przeciwpowodziowej i ich zawodność - przykłady polskie i zagraniczne, Instytut Meteorologii i Gospodarki Wodnej*, Warszawa, <http://www.tnz.most.org.pl/dokumenty/publ/psopp/imgw_w3.htm> [accessed: 1.08.2018].

91 Only 10 % of natural flood plains remained in the Rhine valley. Additionally, the loss of the marshy forests considerably decreased water retention ability of the terrain. (Wawręty and Żelaziński 2007).

92 The hydraulic engineering on a large scale also developed on the Rhine tributaries: the Neckar, the Main and the Moselle and others.

93 Disastrous floods on the Rhine occurred in 1846, 1926, 1983 as well as 1993 (December) and 1995 (January); the last floods were caused by torrential rains in the Rhine Basin and also human activity (engineering of the tributaries, developing riparian terrains). Source: *Encyklopedia* PWN <https://encyklopedia.pwn.pl/haslo/Ren;3967052.html> [accessed: 30.07.2018].

1993 and 1955 (13 months apart!) undermined the effectiveness of the technical flood precautions on the Rhine and drew attention to the search for alternative methods of reducing negative consequences of floods; first of all – returning the space, formerly taken due to the regulation, to the rivers.

Also, the environmental consequences of the river engineering as well as industrialisation and urbanisation of the Rhine Basin proved to be catastrophic. Big cities and industrial plants, especially in the Ruhr district, were the source of strong contamination, while the barriers located in the Rhine riverbed (dams, dykes, weirs and hydropower stations) destroyed the natural migration conditions for fish and other life forms[94]. The embankments, as the basic method of flood precautions, not only caused changes in the flow regime and modifications in the riverbed morphology but also separated the river from its flood plain, destroying the marshy forests and impoverishing water and water-dependent ecosystems. The level of pollution and biological degradation of the river prompted the countries located by the Rhine to cooperate within the International Commission for the Protection of the Rhine, founded in 1950, whose actions gradually led to the improvement of the ecological condition of the river as well as shifts in the environmental law in Europe.

The Danube regulation in Vienna

Large rivers cause enormous problems, especially in sizeable cities. The regulation of the Danube[95] in Vienna is frequently presented as a model example of water engineering achievement, owing to which the city subjugated the problematic river.

The Danube is one of the oldest and the most significant European navigation routes between the east and the west of Europe while Vienna – the most important city by the Danube along with Belgrade and Budapest. The river neighboured

---

94 As a result of damming the river, "many species living in the Rhine died out (e.g. the population of the Atlantic salmon, which in 1870 amounted to 280,000, was reduced to zero in 1950). Also the industrial development contributed – organic waste, heavy metals, hydrocarbon and pesticides contamination. Consequently, life in the Rhine ceased to exist, while the river itself became the sewer of Europe". Source: *Droga do sukcesu*, 13.02. 2003, <http://wiadomosci.onet.pl/kiosk/droga-do-sukcesu/zwxlj> [accessed: 30.07.2018].

95 The Danube, the second largest European river, after the Volga, and one of the most important water routes, was connected with the Rhine in 1933 (the Rhine–the Main–the Danube). Its length: 2,850 km, the basin surface area 817,000 km^2. Source: *Encyklopedia PWN* <https://encyklopedia.pwn.pl/haslo/Dunaj;3894861.html> [4.08.2018].

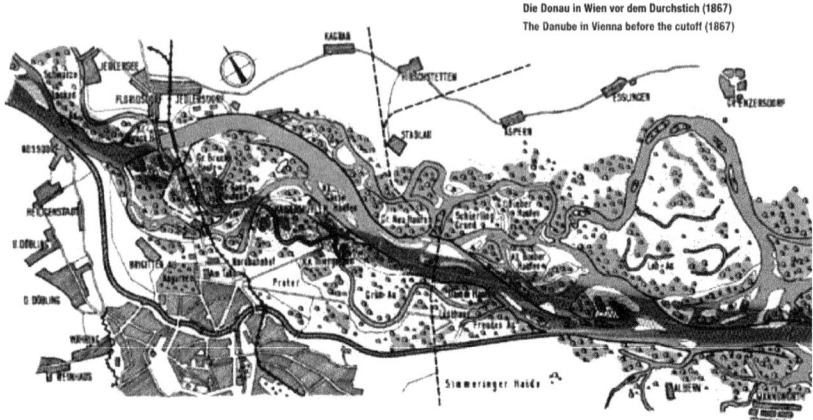

Fig. 2.12: The Danube valley in Vienna in 1867, before the main engineering works. The dark line – the main course of the Danube in accordance with the planned regulation, source: (Donauhochwasserschutz 2017) <https://www.wien.gv.at/umwelt/gewaesser/pdf/donau-hochwasserschutz-2017.pdf> [accessed: 4.08.3018].

the city with its meander tributaries, the main of which, so called the Wiener Arm, provided the waterway access to the city. As a result of the changes in the riverbed, initiated in the 14th century, the Danube gradually moved away from the city and its inlets became silted up and shallow, which hindered the use of the shipping routes and caused damaging floods. Despite the engineering works undertaken in the 16th century, the Danube again moved its course away from Vienna in the years from 1565 to 1566 (Hohensinner 2012). Having failed to restore the main course in the 17th century, the city authorities accepted the situation and focused their efforts on maintaining the navigability of the Wiener Arm (currently Donaukanal) and developing the city fortifications.

In the years from 1775 to 1792, entrenchments and embankments were built with the aim of flood protection, which allowed to expand the development in the former flood plains (Fig. 2.12) but which, simultaneously, changed the dynamics of the flow dramatically (Hohensinner 2012). In the 18th and 19th centuries Vienna, like most of European cities, was undergoing the process of intensive development[96] but the extent of the Danube tides, reaching up to 5 km,

---

96 The capital of the Austro-Hungarian Empire experienced similar processes as other European countries: urbanisation, industrialisation and the increase of the bourgeoisie significance, which influenced the shape of the town planning and the architecture of

did not let the city approach the main river course. The sequence of fierce floods in the first part of the 19th century revealed the weaknesses of the precautions and the power of the element.

In the years from **1870 to 1875 a project on extensive engineering of the Vienna section of the Danube was performed**[97] (Fig. 2.12). The five tributaries of the Danube were reduced to one main riverbed, 280 m wide, and two smaller tributaries. On the left Danube bank, 450 m wide backwater was created along with a flood embankment, which protected the lower located Marchfeld area in Vienna suburbs. On the right Danube bank, higher situated streets, Engerthstrasse and Wehlistrasse, served the function of flood protection. The Donaukanal – the remnant of the former Wiener Arm – flowing through the city, gave a sense of connection with the Danube, whose main course had been straightened and separated from the city. As a result of the engineering in 1875, the achieved level of the precautions and the riverbed stabilisation enabled unrestrained navigation and further urban development but the landscape of the Danube valley was drastically impoverished.

A significant role in the process of water-spatial metamorphose of Vienna is attributed to **Otto Wagner** (1841–1918) – the author of Vienna general retrofitting plan (1893). Wagner's achievements, in addition to the exquisite examples of modernistic architecture, also included the redevelopment of the Stuben Viertel district and the waterfront of the Wienfluss River (cf.: Fig. 2.13) which was a sewage collector flowing through the city centre at that time. Wagner designed its canalization and hiding large sections of the river underground. He was the author of hydraulic engineering structures used for the Danube engineering, i.a. the Nußdorf weir and lock (Bączkowska 2011, pp. 69–82).

The subsequent large floods, which affected Vienna in 1897 and 1899, twenty years after the Danube engineering, proved that the 11,700 m³/s volume of the flood water flow was wrongly estimated (Donauhochwasserschutz … 2017). However, only the catastrophic consequences of the big flood in 1954 forced

---

Vienna. After the uprisings of the Spring of Nations, the fortifications surrounding the city historical part were dismantled, and replaced with a green ring boulevard called Ringstrasse. It consisted of car roads, tram rails and pedestrian areas and, since 1858, was gradually developed with new representative edifices for public and living functions. (More: Kostrzewska 2013, pp. 139–140).

97 The devices applied there were invented by Negrelli, an Austrian engineer. The experiences of a French company called Castor, Couvreux et Hersen, gained during the construction of the Suez Canal, also proved useful (Donauhochwasserschutz … 2017).

Fig. 2.13: The Danube and its tributaries (1–8) in Vienna after the regulation in 1875. (layout by: Friedrich Hauer, based on the hydrographic map of Magistratsabteilung 45 "Wiener Gewässer", city of Vienna)<https://link.springer.com/article/10.1007/s12685-013-0079-x> [accessed: 3.08.2018].

The numbers: tributaries: 1 Waldbach, 2 Schreiberbach, 3 Nesselbach, 4 Arbesbach, 5 Krottenbach, 6 Als, 7 Wien, 8 Liesing, 9 the lowest situated today's Lobau – the remains of the previous flood plains in the south-east outskirts of the city

verification of the flood protection regulations and the development of a new plan of the Danube engineering[98].

In 1969 a project called "Verbesserter Donauhochwasserschutz Wien" (The Improved Flood Control on the Danube for Vienna) was developed. The project received permit under water law in July 1970 and the construction work

---

98 Four forms of flood protection were considered. Finally a multi-functional Variant No 2, based on creating runoffs which decrease flood risk in the floodplain, was chosen (Prof. August Zottl). The flood waters flow was estimated at 14,000 m^3/s and divided: 8.800 m^3/s for the Danube and 5.200 m^3/s for the New Danube. Source: *Projekty*

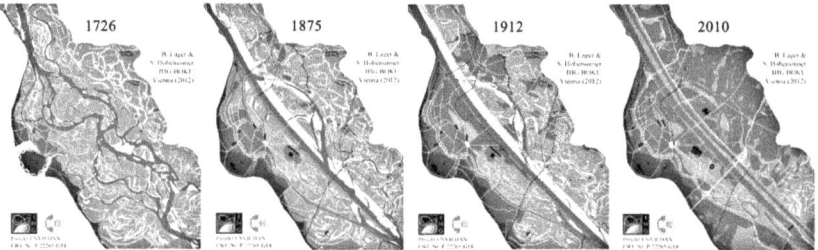

Fig. 2.14: The engineering of the Danube in Vienna. Source: *Danube River in Vienna 1529–2010* (Wiener Donau 1529–2010)<https://www.youtube.com/watch?v=dHERpWgA84Y> [accessed: 3.08.2018] based on the ENVIEDAN research project Nr: P 22265-G18, under the supervision of: Univ. Prof. Ing. Dr.phil. Verena Winiwarter. http://www.umweltgeschichte.uni-klu.ac.at/index,3560,ENVIEDAN.html

lasted from 1972 until 1988 (Donauhochwasserschutz … 2017). In order to unburden the main Danube course, a New Danube canal (Neue Donau) was dug – a retention reservoir for flood waters, which is flow-through only during high river levels[99]. The material from digging the New Danube was used to create embankments of the new Donauinsel island as well as supporting the banks. Donauinsel Island[100], created between the New Danube and the river course regulated in 1875, is currently the most popular leisure area of Vienna's inhabitants (Donauhochwasserschutz … 2017). The Old Danube (Alte Donau), or rather its sections, remained in the form of separated meanders and old river beds which perform a function of city lakes, used as pools and water sports areas

---

ochrony przeciwpowodziowej wzdłuż Dunaju – na przykładzie Wiednia. Ing. Wilfried Fellinger, Magistrat der Stadt Wien, Magistratsabteilung 45 Wiener Gewässer, <www.gewaesser.wien.at> [accessed: 18.08.2018].

99 The New Danube is 21.1 km long and 210 m wide. The runoff is regulated with the use of three dams: the entrance one in Langenzersdorf, the dam near Praterbrückand and the one in Lobau. In this way, two spots which prevent from swelling were created. During the normal water level, the Danube is a stagnant water reservoir. During a flood, the dams are opened which allows to relieve the main course (Donauhochwasserschutz … 2017).

100 The long island (the surface area – 390 ha and the width – varies from 70 m to 210 m) was created from 4 m to 6 m above the previous flooding level and 1 m above the forecast flooding level (Donauhochwasserschutz … 2017).

(Wojnowska-Heciak 2017). Only the main Danube course is used as a transport waterway.

The Vienna flood protection system proved efficient during the big overflow in 2013. The subsequent stages of the Danube regulation are 100 years apart and profoundly differ in the eco-hydrological knowledge level, consideration for the social needs and the awareness of the landscape and the consequences for the nature. However, the comparison of the hydrographical system of the Danube valley before the first big engineering (rich in meander tributaries, islands and old river beds – Fig. 2.12, 2.14)[101], with its current condition, shows **drastic impoverishment of the landscape structure as well as the loss of flood plains with their diverse eco-system.**

The river regulation, undertaken with the aim of protecting from floods and enabling the use of the water routes of the Danube and the Rhine, required agreements and close cooperation of many riparian countries, which was initiated already in the 20th century. The actions continued after World War II, having considerable significance in the process of European integration.

2.4 Water supply-sewage systems and pollution of rivers

Development of cities required improving water-sewage systems effectiveness. In the second part of the 16th century, urban water mains were powered by the use of pumps[102], while the invention of the steam machine (1763) enabled the use of more effective steam pumping stations. The first contemporary water mains systems in Europe were not provided with sewage canals. When they started to appear in the second half of the 20th century, their aim was only sewage discharge but not treatment. Since ancient times until the 20th century sewage discharge reminded cleaning the Augean stables[103] as rivers were constantly the collectors of untreated sewage.

---

101 As part of an interdisciplinary project on the environmental history of the Viennese Danube, the past river landscape was reconstructed. The resulting maps of the Danube floodplain from 1529 to 2010 provide a solid basis for inter-preting the environmental conditions for Vienna's urban development (Hohensinner et al.2013).

102 The first water mains, constructed in 1548 in Augsburg, used water energy to pump water, while in 1582 for the first time in London Peter Morice applied the tidal power of the sea in order to raise the water level in the pipes (using a water wheel working in both directions) [Żukow–Karczewski, n.d.].

103 Cleaning the Augean stables in one day was the fifth out of twelve Heracles' labours which he had to perform for Eurystheus. Heracles channelled the Alpheus or Peneus

The river pollution problem was not apparent in the Medieval Europe as the settlements were still scattered and small towns did not threaten the stability of the regional ecological structures. The situation got out of control along with the intensification of the development and industrialisation of cities. In the 19th century **London** became the biggest metropolis of Europe where 250 tons of municipal sewage and industrial contaminants were discharged into the Thames daily from factories using large amount of water e.g. paper factories, tanneries, dye-works and breweries (Ackroyd 2011)[104]. The rivers pollution caused regular epidemics outbreaks and ecological disasters such as the Great Stink in London in 1858, which motivated the city authorities to have water-sewages system constructed. The works performed from 1869 to 1874 were supervised by engineer Joseph Bazalgette (cf. Chap. 4.3).

All European cities were confronted with similar problems. The first project of a planned sewage system was developed by an English engineer, **William Lindley,** in 1843 for the city of Hamburg[105]. While his son, William Heerlein Lindley, supervised the construction works of the sewage system in Warsaw (in 1876), Włocławek and Prague (an underground sewage treatment system).

The period of the Industrial Revolution not only brought the environmental threats but also technological progress in the fields of sanitary engineering, water purification methods, and sewage treatment. Already in 1829, James Simpson applied sand filters in London water mains systems (Chelsea). Since 1857, the method of water purification with ferrous hydroxide had become popular, and because of the typhoid epidemic in Italy (1896) and England (1897 in Maidstone), water purification by chlorination was implemented. The method appeared so effective that the water mains began to be provided with fixed installations for water chlorination. In the 19th century the first sewage treatment plants

---

river (in some myth versions – both) into the stables. The flowing waters cleaned the stables before dusk.

104 In 1849 city cleaning committee proudly announced that by that time over 465 km sewage canals had been cleaned, removing over 60,000 m³ of sewage which was discharged into the Thames. Source: Michelle Allen: Good Intentions, Unexpected Consequences: Thames Pollution of and The Great Stink of 1858. The Victorian Web. <http://www.victorianweb.org/science/health/thames1.html> [accessed: 2017-08-17].

105 William Lindley – a British engineer – along with his sons, designed and built railway lines, sewage and mains systems for about 30 European cities, i.a. in Warsaw, Hamburg, Basel and Sankt Petersburg. His sewage systems included first underground sewage canals of the early modern period in Europe. Source: <https://en.wikipedia.org/wiki/William_Lindley> [accessed: 14.07.2018]

began to appear along with municipal water supply plants responsible for the quality and quantity of water supplied to the inhabitants as well as collecting and treating sewage. The commonly used method of preliminary sewage treatment was the method of fermentation in clarification plants and spreading the sewage in so called septic drain fields, used i.a. in Berlin[106] and Brandenburg, Gdańsk, Wrocław or Legnica. Many of such irrigation treatment plants were still in working order till the 1990s.

The design of the municipal water supply-sewage systems was based on the principle of water extraction in the upper reaches of the river and sewage discharge in the river lower reaches (below the city). Also the industrial plants were situated below water intakes. The environmental consequences of the rivers pollution were not considered though. The change in the approach to the water supply-sewage systems and tightening the regulations occurred only when the condition of the surface waters became disastrous along the entire length of many European watercourses, as a consequence of exceeding their self-regeneration capability and the impoverishment of the natural environment of the river valleys due to the rivers engineering and canalisation.

2.5 Canalisation of urban watercourses and drainage of swamplands

Small rivers, creeks and streams used to supply municipal moats and provide water and energy to the inhabitants. . Popularisation of the water wheel facilitate development of the urban hydrographical system, enriching it with numerous canals – so called mill streams – which, after damming water, created lakes and ponds, while the energy supplied by them was used in various branches of craft and industry. The situation radically changed along with the increase in the watercourses contamination. The construction of the water mains solved the problem of supplying water but the urban swampland, contaminated with the sewage flowing from the city, emitted foul odour which raised social objections. Discovering the link between the regular epidemics outbreaks and the water contamination resulted in **"declaring war" against all the wetlands in the city.**

---

106 An innovative project which prevented discharging the sewage into the Spree was developed by James Hobrecht. Decentralised sewage system of Berlin consisted in the division of the city into twelve districts, each provided with an independent sewage system and a pumping station, which allowed to pump the sewage to septic drain fields beyond the city (Gray 2014).

The rivers and the canals, which were useful for navigation and industry, constituted the framework of the urban structures, while the small rivers, streams, old riverbeds and moats were successively canalised under the ground or backfilled. Swamplands and stagnant backwaters were considered urban nuisance becoming hotbeds for mosquitoes and diseases as well as dangerous and troublesome wasteland which occupied the valuable city space. No wonder, reclamation, draining and "regaining" the wetlands[107] for the needs of housing, industry and communication was, by all means, the desirable objective of urban investments. The drained swamplands made space for new districts[108] or parks. In non-urban areas the wetlands were commonly reclaimed for agricultural demands. The consequences of the wetlands liquidation can be compared to squeezing a sponge which previously used to absorb the excess of overflow waters. In the 19th century the problem of wetlands disappearance became a global issue[109].

At the turn of the 19th century the majority of small watercourses disappeared from the landscape of European cities. As a result of gradual canalisation and backfilling of formerly plentiful streams, leats and moats, the hydrographical structure of Poznań - once "a city on five islands" created by the numerous Warta tributaries – has been strongly impoverished (Fig. 2.15, Kaniecki 2004, Januchta-Szostak and Biskupski 2014). The regulation process of the river led to straightening its course and resulted in backfilling the Chwaliszewo Meander.

The degradation extent can be well illustrated by the example of Łódź – a city which was founded and quickly developed in the 19th century owing to the textile industry which used water from its 19 creeks and streams. The dense hydrographical system supplied the factories and collected industrial and municipal sewage of Łódź[110]. In view of the growing pollution level as well as acrid

---

107 Many cities were founded in the areas where the swamplands were drained (i.a. Amsterdam and Sankt Petersburg or Dolne Miasto in Gdańsk) and the reclamation works were mainly performed by the Dutch settlers.
108 E.g. Brooklyn in New York, Saska Kępa in Warsaw, St. James Park in London in the 18th century and Central Park in New York in mid-19th century (project by Frederic Law Olmsted and Calvert Vaux).
109 The problem was addressed in 1971 by the Ramsar Convention concerning the protection of wetlands.
110 At the beginning of the 1930s the Polish leading columnist, Zygmunt Nowakowski, wrote: "Łódź ground is fertilised by a long row of gutters which cross the city in a random, spontaneous way, making the Polish Manchester a serious rival for Venice. Since the entire city is one large 'Canale Grande'" (Nowakowski 1931).

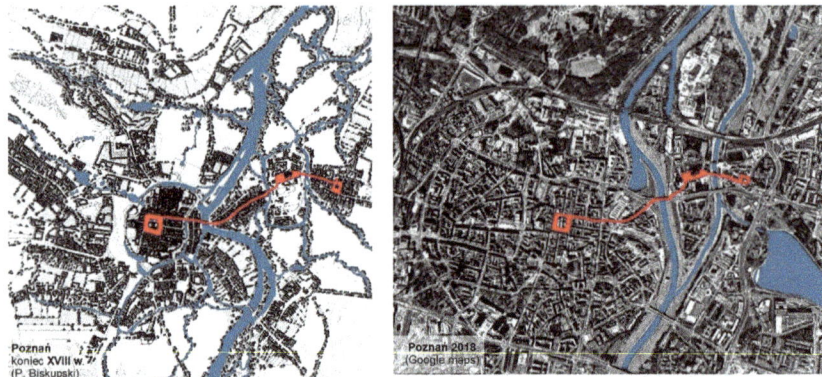

Fig. 2.15: A comparative profile of Poznań hydrographical structure at the end of the 18th century and a contemporary picture. The marks: the red line – the route connecting the Śródka and Poznań markets; the blue areas – hydrographical structures (developed by A. Januchta-Szostak based on historical map developed by P. Biskupski – on the left and contemporary Google map – on the right).

stink, considerable lengths of the rivers of Łódź were canalised and hid underground (Fig 2.16)[111]. As early as before 1913 some sections of the Bałutka, the Stoczanka and the Dąbrówka disappeared from the surface. Between 1913 and 1941 sections of the Łódka, the Jasień and the Karolewka were transformed into underground canals, while the Lamus stream almost completely dried up (Stolińska 2018). In the 20th century the majority of small watercourses flowing through the city centre were hidden underground, transformed into sewage canals and removed from the urban space and, over time, from human memory.

In big cities not only streams but also sizeable rivers were canalised underground, like the Nieglinnaja in Moscow[112] and the Pełtew in Lwów[113] (Fig 2.17),

---

111 The map of the rivers of Łódź was based on the materials facilitated by the Public Record Office in Łódź, Urban Town Planning Studio as well as archival photos of Waterworks and Sewage Utility. It helps to discover where the watercourses flowed at the time of the industrial prosperity of the city. Two former mills and spots of "the tourist trail of the Łódź rivers", created by ZWIK, are also marked on the map. More: ZWIK Łódź <http://www.zwik.lodz.pl/mapa-lodzkich-rzek/> [accessed: 29.07.2018].
112 In the past, the Nieglinnaja river constituted a section of the Kremlin's moat and flows under the very centre of Moscow (Kowalczak 2015, p. 84).
113 The 70-km-long Peltew river (Ukrainian Полтва Połtwa) is the left tributary of the Bug River, flowing in Lwów oblast of Ukraine at its whole length. The river created

Canalisation of urban watercourses and drainage of swamplands 73

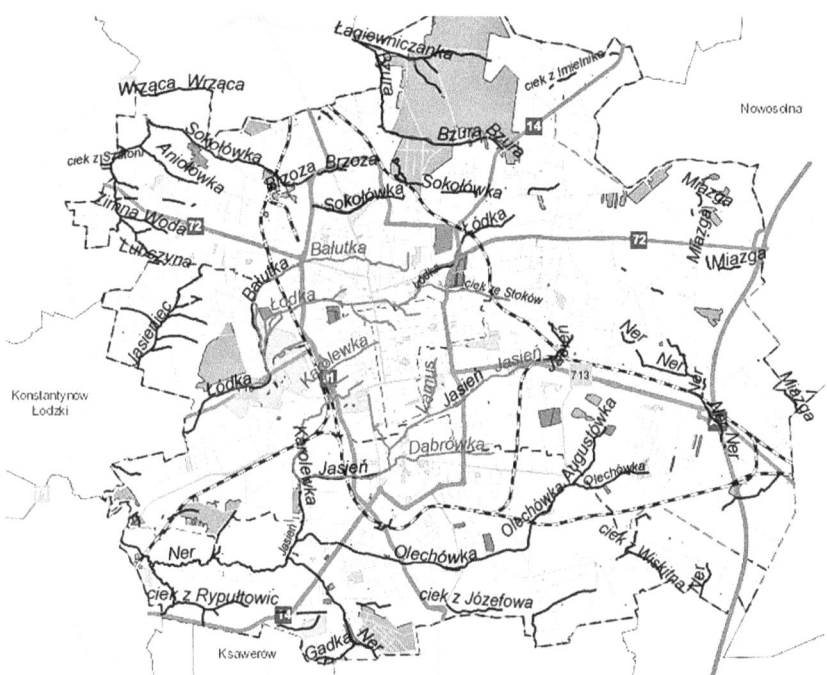

Fig. 2.16: The hydrographical system of Łódź: the rivers entirely or partially canalised (coloured dark blue) and non-existent sections of rivers (coloured light blue). Based on the study by M. Stolińska (2018) and source maps, source: http://mapa.lodz.pl/portal/apps/webappviewer/index.html?id=7489987eb60a4369814e53d49cc58ffc&extent=19.317,51.7126,19.6459,51.8507 [accessed: 1.06.2018].

which, at the end of the 19th century, was covered in the midtown section and included in the urban sewage system as the main collector[114]. Their fate was

---

swampy back rivers and at the time of foundation (ca.1250) and development of the medieval Lwów, it served as a city moat and a navigational river. The decision was taken because of the malarial threat for the city. Source: <https://pl.wikipedia.org/wiki/Pe%C5%82tew> [accessed: 28.07.2018]. In 1887, the Pełtew was transformed into an underground canal and disappeared from the landscape of Lwów. More: Petryshyn, Sosnova 2016, pp. 113–114.

114 The initial length of the underground sewage system, whose part was the Pełtew in 1870, amounted to 15 km. Just before the outbreak of World War II, in 1939, it was extended to 150 km. Source: Paweł Łacheta, 31.08.2013, Lwowskie podziemia <https://pawellacheta.com/2013/08/31/lwowskie-podziemia/> [accessed: 2.08.2018].

Fig. 2.17: The Pełtew River buried in the underground of Lwów. Photo: Vladislav Vozniuck, source: <https://general-kosmosa.livejournal.com/61009.html> [accessed: 1.08.2018].

Fig. 2.18: The view of the development along the La Bièvre River, the tributary of the Seine in Paris (la rue des Gobelins) before the redevelopment by Haussmann in the second half of the 19th century, source: Crossman, 2013.

shared by the la Bièvre River in Paris (Fig. 2.18), the Wienfluss in Vienna, 21 London rivers and streams as well as thousands of small watercourses in other European cities such as Potok Służewiecki in Warsaw, sections of the Rudawa and the Prądnik in Kraków or the Bogdanka and the Wierzbak in Poznań. By the 1970s, hiding the contaminated creeks under the ground had been a common "reparative" method; regrettably, such measures are still taken in the 21st century like in the case of the Rawa River in Chorzów and Świętochłowice[115], covered in 2010, which was financed by the European Union funds.

---

115 In 2010 within the framework of the task No 2 "Covering the Rawa riverbed along with the sewage canals", financed by the EU funds, about a 5 km long section of the Rawa riverbed was covered and transformed into a 5 km closed sewage canal. (CHŚPWIK <https://www.chspwik.pl/inwestycje/rawa.html> [accessed: 20.07.2018]).

As a consequence of the actions, **small narrow urban watercourses, especially their midtown sections, lost their biological, hydraulic, landscape and cultural vitality**. The majority of them were canalised in concrete or brick beds due to which they lost their vegetation necessary for self-purification processes. Huge amount of pollution effectively deprived them of **biological** life. The destruction was complemented by covering the streams in the underground collectors. Frequently, they finished their **hydraulic existence** as well – separated from the water supply sources, they just dried up and their beds were backfilled. Their **cultural existence** ceased along with the loss of identity – they were no longer rivers, streams or creeks but rather anonymous sewage collectors which remain neither in the landscape nor on the maps and, gradually, disappear from the inhabitants' memory.

2.6 Beautification of the nature and cities

The Italian Renaissance creators, inspired by the antique knowledge, discovered again the aesthetic and mythological imagery of water. The theoretical foundations for the arrangements, integrating the architecture with its surroundings, based on the ancient patterns were created by: L.B. Alberti (1404–1472), A. Bramante (1444–1514), R. Santi (1483–1520) and G. Romano (1492–1546). The designers again acquired knowledge on architecture, town planning and hydraulic engineering, which not only concerned the ways of water elements arrangements but also the methods of draining and irrigating the terrain, from the works by Archimedes, Aristotle, Vitruvius, Heron from Alexandria. The studies and inventions by Leonardo da Vinci (1452–1519) led to considerable enhancement of the knowledge on i.a. anatomy, construction and architecture, sculpture, philosophy as well as hydraulics and hydrodynamics[116]. The development of hydrology in the 17th century was facilitated by the works of Galileo, Descartes, Castelli (calculating the flow rate) and Kircher (the origin of springs and rivers) as well as the foundation of the Royal Society in London in 1660 and the French Academy of Sciences in Paris in 1666 (Biswas 1978).

Descartes' ideas and his scientific approach to the world's perception also fundamentally influenced philosophy, ethics and economics. As Tomáš Sedláček

---

116 Leonardo's studies on water movement inspired him to design machines which used its power. The majority of his works were performed for Ludovico Sforza. Considering possibilities of canalising the swampy terrain neighbouring the palace, he had an idea of producing a device which would extract the groundwater. The device pumped water with the use of a screw, rotating inside a cylinder (Vassari 1998).

(2012) points out, following V. Mini's thought, limiting the universe, society and the man to mechanical-mathematical dimensions, laid the foundation for the birth of *homo œconomicus* – "the narrowest concept of the man one can imagine" (Mini 1974, p. 24). From the rational man's point of view, which still determines the perception of the world, it seems logical and substantiated to aim at increasingly venturous transformation of the environment, arranging and "improving" the nature in accordance with classical composition principles as well as creating huge spatial complexes of modern period. New fields of knowledge, inventions and skills enabled effective implementation of new ethical, aesthetical and economic beliefs.

Parks and gardens

Unlike rivers, reservoirs in parks and gardens[117] were protected from pollution. In Renaissance gardens of the entire Europe of that time, it was the "Italian school" which dominated, referring to the axial or central arrangements of Roman villas and palaces connected with gardens, while the hydraulic devices and hydraulic engineering inventions were used for decoration and entertainment[118].

French water park arrangements developed from former castle moats. The water canals in the Loire valley served both the purpose of wetland reclamation and beautifying the garden arrangements. The gardens in Fontainebleau and Chenonceaux, redeveloped by Catherine de' Medici after 1560, became the scenery for performances given on water (referring to Roman naumachiae), which initiated the fashion for water axes[119], developed in the Baroque. Water arrangements and technical novelties were followed in Paris (Saint-Germain-en-Laye by Henry IV), Prague (the gardens of Rudolf II), Heidelberg (Hortus Palatinus of Frederic V), London (Richmond Palace) and many other European

---

117 More: Majdecki 1981; Hobhouse 2005; Sosnowski and Wójcik 2008; Zachariasz 2006.
118 Pratolino – "wonders' garden", designed in 1569 by architect Bernardo Buontalenti for Francesco de' Medici near Florence, famous for sophisticated water devices and so called plays of water (giochi d'acqua), (Hobhouse 2005, p. 129).
119 Artificial canals were the main compositional element of French gardens and they reached an impressive size (in Versailles – 1,670 m, in Vaux le Vicomte – 900 m, in Fontainebleau – 1,100 m). The junctions, beginnings or endings of canals were accentuated by fancy geometrical pools which constituted so called water parterre. The gardens were decorated with sculptures and fountains. In order to achieve even better results, cascades, water steps and curtains were constructed at the closing points of view axes as a sign of prestige and technical abilities of the age – increasingly more superb pumps and pressure devices.

cities, while the "naturalism" of the Italian Mannerist gardens strongly influenced the shape of the 18th century landscape parks.

Life conditions in big cities (Paris, London) deteriorated which prompted the aristocracy to found outskirt mansions. In the second part of the 17th century in France, **Andre Le Nôtre**, a royal gardener at Louis XIV's court, initiated the intense development of **French Baroque gardens**. Masterpieces by Andre Le'Notre's include the gardens of Fontainebleau, Vaux-Le-Vicomte, Chantilly, Tuileries in Paris, Marly and the monumental gardens of Versailles[120], which became the pattern for many garden arrangements in Europe. If Louis XIV had been constrained by the environmental impact assessment, apparently the big Versailles arrangements would have never been created. Their construction required flattening and reclamation of 1,600 ha of swampland and forests in order to transform them into geometrical lanes (boskets) of trimmed trees, flowery, water parterres, canals and pools with fountains which were supplied with water from the Seine (60 km) using 257 river pumps powered by 14 waterwheels.

Similar scale of splendour and water arrangements panache could be only found in the big garden of Peter the Great in Sankt Petersburg, called **Peterhof**[121],

---

120 The Gardens of Versailles cover the area of over 8,000 ha and their arrangement is based on a cross plan and accentuated by the Grand Canal (surface 23 ha, length 1,670 m), which was built from 1667–1671, replacing of de Galie stream and draining the wetland of the valley. The arrangement consists of the Petit Trianon and the Grand Trianon. The fan-radial layout of the Grand Trianon features eight axes coming from the Apollo's pool. The Versailles area is scarcely topographically varied and surface waters deficient. In order to supply it with water, it was necessary to install sophisticated, expensive and highly effective hydraulic devices. Le Nôtre developed a method of soil testing as well as the system of extraction, retention and distribution of water. A special intake was built on the Seine in Marly, where huge amount of water was pumped to supply the gardens. A three-stage, several-meter water lifting was used to, subsequently, gravitationally distribute the water with a system of underground pipes to individual fountains which spouted water a few meters in the air. More: Majdecki 1981; Hobhouse 2005; Sosnowski, Wójcik 2008.

121 The construction of Peterhof Grand Palace, which began in 1714 supervised by Alexandre Le Blond – French and later Bartolomeo Rastrelli – Italian, transformed the terrain into a huge arrangement (21 ha) with a water axis facing towards the Gulf of Finland, great cascade and hundreds of fountains. The Upper Gardens were arranged in the French style, while the Lower Gardens – in the English style. They are divided by the Grand Cascade, completed in 1724, decorated with 37 gilded bronze statues, 64 fountains and 142 waterspouts. Some mechanical fountains are interesting examples of water art, including the Sun Fountain in the form of a high column with gilded discs which spout 72 water streams reminding sun rays. The water is channelled to the park

whose construction required draining the swamplands, which was performed with relentless consistence. The role of the huge royal palace-garden arrangements was to reflect the power and the majesty of the absolute monarchy[122], which reckoned neither with the society nor the nature.

The 18th-century **English landscape parks** seem to be much closer to the nature in comparison to the Baroque geometrical compositions. The basic difference between a French and an English garden lay in using the natural beauty of the wilderness, but the irregular English garden plan was supposed to make the nature even more picturesque. Natural landscape, undergoing erosion and plants succession processes appeared too common, inaccessible, wild and difficult to control. Therefore, the random, asymmetrical sequences of parks' interiors were created in accordance with the varied relief, land cover, natural springs, watercourses and water reservoirs. In case of their shortage, the roughness of the ground was artificially modelled, backwaters with islands were created and ponds in irregular shapes were dug in the ground. Instead of geometrical pools and canals, curvy-lined creeks and streams were created. However, the restrained nature required constant care of gardeners. The first designer[123] who fully introduced the new principles to the garden solutions was **William Kent** (1685–1748) – a painter, architect and stage designer. In the Rousham[124] garden, nearby Oxford, Kent used the picturesque Cherwad River valley, with forest slopes, in order to create a vast 17-hectare park arrangement. The aesthetics of English gardens considerably influenced the shape of urban architectural-park compositions. In the second part of the 18th century, numerous parks and gardens in Europe were founded or redeveloped in accordance with the English style, e.g. the Petit Trianon in the Versailles, Monceau park in Paris, Worlitz park in Dessau or the park in Weimar. St. James Park and Regents Park are examples of parks redeveloped at the beginning of the 19th century by John Nash. Both

---

    from Wzgórza Roposzyńskie (hills located 22 km away), it flows down to the White Sea Canal and further to the Gulf of Finland. All the fountains use up to 100,000 m³ of water daily (Gööck 1994, pp. 148–151).

122 Many European absolute monarchies legitimated their reign by so called divine right of kings. The representative of the concept in England was Sir Robert Filmer, the author of the treaty titled *Patriarcha* (1680).

123 The first realisations in accordance with the garden arrangements appeared before in lateral segments of gardens, e.g. Twickenham realised about 1714 by A. Pope near London. The garden in Chiswick was arranged in the same way.

124 The formal garden created by Bridgeman was redeveloped by Kent in the years of 1788–1741.

of them were enriched with water elements, and Nash was also the author of the Regent Canal course.

City densification

The large free-standing space arrangements, like palace and garden complexes in Versailles (1634–1697) and Peterhof (1703), or the city of Karlsruhe (1710) followed the compositional transformations of Rome, initiated during the short pontificate (1585–1590) of Sixtus V, later called "the first modern town planner"(Böhm 2006, p. 32). The great Baroque redevelopment of Rome[125] enriched the Eternal City not only with architectural masterpieces but first of all with new, axis town planning arrangements, concatenated with representative squares which became the pattern for other European cities. Water beauty manifested itself mostly through fountains decorating Roman squares (Januchta-Szostak 2017a), while the Tiber was becoming an urban gutter.

The development of the majority of old European cities resembled making "snow balls" as they grew by increasing compact development which blocked the access to open terrains. Parks and gardens were reserved for the privileged social groups; however, in the 17th and the 18th centuries, the significance and the pressure of bourgeoisie increased so much that they began to be facilitated to the public use. Already in the 17th century in Paris, the first gardens opened for the inhabitants: Tuileries and Luksemburg, founded by Marie de' Medici, and in London – Hyde Park which became a public space in 1637. City plazas[126], public promenades and boulevards began to appear and the useless medieval walls and city moats were replaced by parks (e.g. green strolling area in Cracow). Terrains in the vicinity of big rivers were location areas of representative edifices (e.g. in Paris or Budapest) or harbour, craft, and later industrial facilities (e.g. Rotterdam, Amsterdam, London[127]).

Relations between Paris and the Seine River were a model for many European cities at that time (Fig. 2.19). The Seine constituted the compositional city axis, which was lined with boulevards and green spaces (Tuileries gardens) as well as

---

125 For the following 150 years the redevelopment was continued by Antonio Bramante, Michelangelo, Carlo Maderno, Lorenzo Bernini and others.
126 Covent Garden, built for the earl of Bedford, in accordance with the project by Inigo Jones, constitutes an example of new type of housing development integrated with common space; but only St. James Square, as one of the first, was arranged for leisure purposes in 1726 (Bartkowicz 1985, p. 159).
127 London example was described in more detail and illustrated in Chap. 4.3.

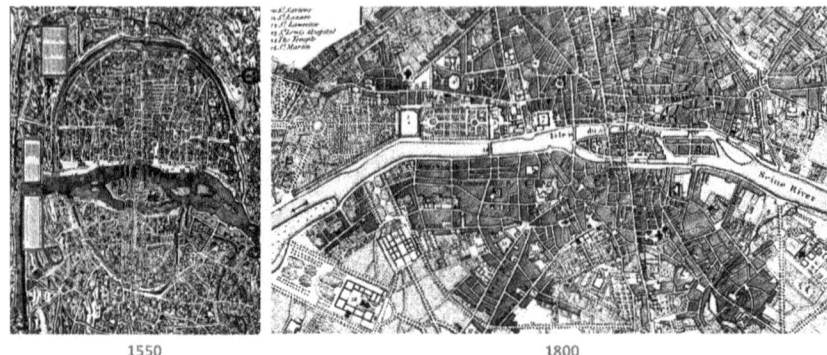

1550   1800

Fig. 2.19: A comparative profile of Paris plan from the mid-16th century[128] (on the left) and part of a plan from the turn of the 18th century[129] (on the right).

representative edifices like the Louvre, Les Invalides (1671–1676) – its axis towards the river, or the Army School complex (1752–1772), connected with the Seine by the Champ the Mars. **The Great Boulevards** founded by Louis XIV became the first representative riverine public space and the main leisure area for the Parisians. Despite the densification of the development and shortage of greenery, the very contact with the open river space, even if limited by the frame of the stony embankments, offered respite to the inhabitants of the overpopulated city.

In the 18th and the 19th centuries, the urban areas were expanding and becoming increasingly congested due to the advancement in industry, which offered posts in factories. The land prices were growing, while the living conditions were deteriorating dramatically[130]. Only the outbursts of desperation of the poor social classes, revealed during the French Revolution (1789–99) and the Revolutions of 1848, impelled the decision makers to have the public spaces of

---

128 The plan of Paris, source: *Amazing Maps of Medieval Cities*, <http://dailyinfographics.eu/amazing-maps-of-medieval-cities/> [accessed: 7.11.2018].
129 The plan of the city published by John Stockdale in: *A Description of the Empire of Germany, Holland, the Netherlands* ..., 1800. Source: <http://www.ancestryimages.com/proddetail.php?prod=f5899> [accessed: 7.11.2018].
130 "According to the data presented by Philip Freriks in *Południk Paryża*, the city gradually lost control over the rapid growth. The population grew from 0.5 m in 1815 to 1 m in 1848 and nearly 2 m in 1870. For example, Île de la Cité was inhabited by almost 15,000 people and the total Paris population increased from 2 to 7 % "(Seroka 2012).

Beautification of the nature and cities 81

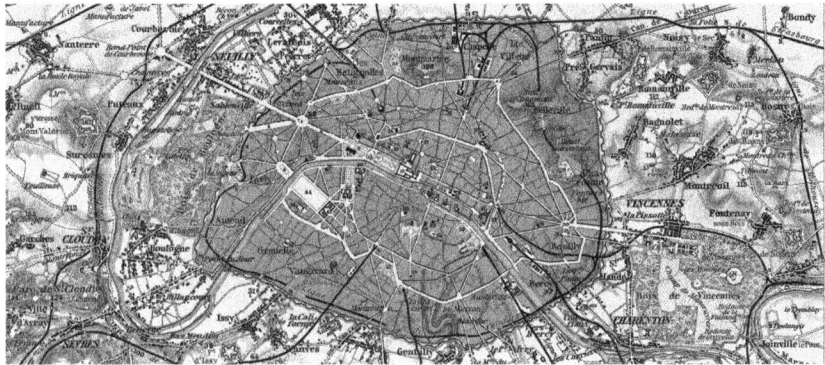

Fig. 2.20: Part of Paris plan from 1875, after the redevelopment by E. Haussman[131].

big cities transformed. The redevelopment of Paris in the 19th century, however, did not aim to enhance the poorest inhabitants' living conditions but rather to dispose of poverty from the city centre and regain the control of the Paris space.

The first stage of the development in the years of 1833–1848 was supervised by **Claude Rambuteau**[132], who started from planting trees, reinforcing the embankments of the Seine and building two bridges. However, only the big redevelopment of Paris from 1852 till 1870, called Haussmann's renovation of Paris, changed the image of the city (Fig. 2.20). Napoleon Bonaparte, inspired by the redevelopment of London, planned to transform Paris into a modern city, which would measure up to the functionality and aesthetics of the capital of England.

The subsequent prefect of Paris, baron **Georges Eugène Haussmann**, charted a new system of communication arteries[133] in the form of straight 20–40 m boulevards with regular facades and representative development of public

---

131 Author: Adolf Stieler, title: *Paris und Umgebung*. Frankreich In 4 Blattern. Von C. Vogel. Maassstab=1:1.500.00 0 der naturlichen Lange. Blatt 3. Gestochen: Schrift von H. Eberhardt. Terrain von W. Weiler. Gotha: Justus Perthes. 1873. David Rumsey Historical Map Collection, Source: <https://commons.wikimedia.org/wiki/File:Adolf_Stieler,_Paris_und_Umgebung,_1875_-_David_Rumsey.jpg> [accessed: 7.11.2018]
132 The contemporary prefect of the Seine department, the Earl Claude Philibert Barthelot de Rambuteau, also increased the number of city fountains supplying drinking water (from 100 in 1830 to almost 2,000 in 1848) and had the city street lighting gas system installed. He had also a wide boulevard constructed along the trade district of Les Halles. More: Walter 1975; Freriks 2007; Seroka 2012.
133 Haussmann began the reorganisation of the city with the construction of so called grande croisée – a huge cross along the axes east-west (rue de Rivoli and rue

edifices[134] and stations. He also had water-sewage system constructed[135] and created urban leisure areas: parks (e.g. Buttes-Chaumont or Monceau), garden and municipal forests (Bois de Boulogne – 873 ha and Bois de Vincennes – 920 ha).

As a result of the transformations, the greenery surface in Paris in 1870 increased 22-fold, which gave 11 m^2 of park greenery per an inhabitant (Bartkowicz 1985, p. 22). Rearrangement of the city also involved the small contaminated rivers and streams like the la Bièvre river – a 36 km tributary of the Seine, flowing through the Paris centre, which was integrated with the city sewage system and hid underground (Fig. 2.18).

The transformations of cities during the Industrial Revolution not only caused deterioration of the living conditions in cities but also degradation of their leisure potential (Bartkowicz 1985, p. 13). First of all, already in the 17th century, cites **rapidly** began to **expand territorially**; by the 21st century, especially the areas of the capitals of colonial empires had increased several-fold, which resulted in surrounding them with "endless" suburbs. Secondly, **the densification of the inner city structure** led to disappearance of greenery not only in the public but also half-private space – inside the quarters, which reduced leisure opportunities in the city. Finally, **not only the urbanised environment but also the cities' surroundings succumbed to degradation**. Due to forest logging for fuel and industrial needs, the suburban areas were devastated and the rivers – heavily contaminated, which impoverished their attractiveness as leisure terrains. Regardless of the background and genesis, urbanization processes in Europe in the 19th and 20th centuries (Bronski and Szpak 2002) had a strong impact on the scale of environmental degradation.

The purpose of the repair actions undertaken in the 19th-century European cities was to create opportunities of further development and to enhance the living conditions. J. Bazalgette in London, G.E. Haussmann in Paris, W. Lindley in Germany and Central Europe or O. Wagner in Vienna performed their projects in accordance with accessible engineering knowledge as well as the system of values prevailing in that place at that time. There was no consideration for the environmental consequences of the urbanisation, industrialisation, much less for the rivers engineering.

Saint-Antoine) and north-south (de Sébastopol and Saint-Michel boulevards), connecting Gare de l'Est (Station of the East) and Gare du Nord (Station of the North) with straight roads.

134 E.g. the Opera designed by Ch Garnier or Les Halles – by W. Baltard in 1863.
135 Haussmann left approx. 272 km of water mains and over 600 km of sewage canals (Seroka 2012).

2.7 The scale of the changes and the price of the progress

Severe floods have been occurring since the dawn of time, however, the scale of **urbanisation** in the 19th and the 20th centuries as well as the extent of the transformations of the basins and river valleys have compounded their catastrophic consequences (Kundzewicz 2012). The changes of land use, as a result of spatial expansion of cities and draining the swamplands, increased the rate of basin runoff while rivers regulation narrowed the valleys which raised the flow rate and flood water levels.

At the time of the Industrial Revolution, the direct **access to waterways** determined intense economic development of cities but the excessive development densification along the rivers led to **flood risk surge**. Unlike historical city centres and representative districts, which were located in the areas above flood plains, in the 19th century the zones of industrial development[136] and the working-class districts were situated in the direct river vicinity on heightened waterfronts or just behind flood embankments.

The need for protection of the urbanised flood plains entailed the necessity of heightening the river embankments, which made illusive sense of safety, encouraging yet another investment in the embankment-protected areas. Breaking them could arise severe consequences, thus the technical precautions were reinforced which in turn eventuated in **a vicious circle of flood protection** (Bobiński and Żelaziński 1996).

As a consequence of the regulation some river courses were shortened even by a few hundred kilometres[137], which compounded the destructive power of the flood overflows. The technological progress made it possible, however, to create increasingly more advanced flood protection systems, including not only the embankments but also retention reservoirs, flooding polders, flood relief canals, barriers and dams. All the undertakings were located in the river valleys where the flood consequences were most acute while the urban basin areas had been swiftly drained (Kowalczak 2011).

---

136 The development of the Ruhr district, Manchester or London can be illustrative examples of the phenomenon. The use of the riparian areas for industrial, transport or energy technology purposes was suggested by Tony Garnier in his theoretical concept of industrial city (reprint: Garnier 1989), partially accomplished in Lyon at the beg. of the 20th century.

137 As a result of the engineering of the Odra river, its length was shortened by 160 km; from its original length of 1,020 km to 860 km (Bobiński and Żelaziński 2015). And the length of the Rhine was reduced by 81 km.

Tab. 2.1: The conquest period – the most critical changes within the river valleys and city catchment areas as well as their consequences – analysis of RGB urban structures

The transformations of river valleys and city catchment areas	The consequences of changes: analysis of RGB urban structures		
	Cultural-economic	Nature	Hydro-morphological
	R	G	B
River engineering straightening the course (cutting off the meanders), reinforcing the banks; constructing navigational canals, damming rivers	The improvement of navigation conditions; extension of the waterways system; development of water transport and trade; water energy use; erosion protection; transformation of river valleys landscape.	The impoverishment of riparian wildlife; fragmentation of ecosystems – breaking the continuity of water and valley ecological corridors; deterioration of the conditions of wildlife migration; limiting self-purification capability and biological recovery.	The improvement of big waters flow; stabilisation of river beds; shortening watercourses, increase in water runoff rate; decreasing the riverbed and valleys retention; changes in hydro-morphology of watercourses; increasing water erosion.
Flood embankments	Temporary flood protection; new investment areas reclamation; development of flood plains – increase in flood risk; impoverishment of river valleys; cost of the embankments maintenance and retrofitting.	Separating the flood plains; transformations and degradation of water-dependant ecosystems, e.g. riparian forests; maintenance of the embankment and inner embankments – logging and clearing the bushes, exterminating beavers and burrowing animals.	Drastic valley retention decrease; runoff rate and flood wave rise; increase of the destructive overflow power; decline of the watercourses self-purification capability.
Uncontrolled spatial expansion of cities	Changes in developing and using the suburban areas and flood plains and spatial expansion, growth of residential and housing development; communication problems; deterioration of leisure conditions;	Degradation of city surrounding areas: logging forests, draining the swamplands and degradation of wetland meadows; impoverishment of the wildlife; fragmentation of ecosystems;	Changes in the profile of catchment areas and dynamics of city watercourses – increase in water runoff rate; water pollution level exceeding the regeneration capability of the environment;

Tab. 2.1: Continued

The transformations of river valleys and city catchment areas	The consequences of changes: analysis of RGB urban structures		
	Cultural-economic	Nature	Hydro-morphological
	R	G	B
Densification of inner structures, sealing the cities	Increase in development densification and population; deterioration of life conditions and leisure opportunities;	Degradation of green areas; reduction of biologically active areas – sealing; problems with airing cities;	Increase in rainwater runoff rate; growth of flow dynamics in small watercourses; urban floods;
Industrialisation, industrial development along river valleys	Using rivers in technological and transport processes; economic growth, job vacancies, separating the city from the river by the industrial zones;	Destroying river valley ecosystems; breaking the continuity of eco-corridors; water, soil and air contamination;	Considerable water extraction and discharge of the industrial sewage including chemical, biological and thermal pollution; instabilities in flow dynamics;
Water – sewage systems of big cities (without sewage treatment plants)	Discharging city sewage; epidemiological risk; loss of ecosystem functions including the opportunities of swimming and resting;	Destruction of water and water-dependant ecosystems; environmental pollution; degradation of river landscape;	Increasing water extraction; discharging not purified sewage – catastrophic river contamination;
Canalisation of small watercourses, reclamation and draining the wetlands	Liquidation of acrid smell and epidemiological risk; gaining new investment and farmland areas; loss of ecosystem services;	Extinction of biological life of watercourses; liquidation of valley green structures and eco-corridors; degradation of wetland ecosystems; urban heat islands;	Dramatic changes in watercourses hydro-morphology; impoverishment of hydro-graphic system; lack of surface and rainwater retention possibility; lack of self-purifying capability of watercourses;

(*continued on next page*)

Tab. 2.1: Continued

The transformations of river valleys and city catchment areas	The consequences of changes: analysis of RGB urban structures		
	Cultural-economic	Nature	Hydro-morphological
	R	G	B
"Beautifying and civilising" the nature	The development of landscape architecture; shaping public space of waterfronts – first boulevards and promenades; implementing greenery structures in cities;	Creating artificial landscapes of parks and gardens – introducing alien species at the expense of local ecosystems;	Changes in hydro-morphology of watercourses; draining swamplands, retention decrease; considerable water extraction;

Also **urban floods**[138] caused by fast runoff of precipitation whose volume exceeded the capacity of urban sewerage systems, proved to be a new dangerous phenomenon of the 20th century. Due to the spatial expansion and rising densification of inner city structures, the runoff conditions in city catchment areas radically changed (Kowalczak 2011, 2015). Densely developed, cramped districts, lack of greenery, draining the wetlands and canalising or liquidating small watercourses[139] contributed to considerable limitation of retention and infiltration capacity of precipitation water. In the 21st century, the extremely drained cities are hardly prepared to face new extreme hydro-meteorological phenomena caused by global climate change. The scale and consequences of the transformations within the river valleys and city catchment areas are illustrated by Tab. 2.1.

The disastrous level of rivers pollution in the 1960–1970s resulted not only from lack of requirements and efficient purification systems of municipal and industrial sewage but also from sealing considerable basin areas and separating

---

138 Piotr Kowalczak (2011, p. 169–170) gives an example of Athens, which experienced dramatic urban floods as a result of rapid urbanisation at the beg. of the 20th century. For the last 120 years, the floods in the city have caused 180 casualties

139 According to cartographic analyses from the end of the 19th century till 1988, within the area of the present time Kliffiss basin in Athens, 113 km of watercourses, which means 20 % of natural drainage, ceased to exist (Kowalczak 2011, p. 170).

the rivers from their natural "sewage treatment plants" i.e. the swamplands and riparian ecosystems in the flood plains. Ever since 1900 over 64 % of wetlands[140] have disappeared globally so far. Only from 2009 to 2016, as a consequence of anthropopression, 33 % of water-swamp areas disappeared in the world and in Europe as much as 45 % (Hu et al 2017). At the end of the 1960s, the contamination of the environment began to constitute a barrier to the socio-economic development (Chmielewski 1996). On a global scale the conquest began a new age in the history of the Earth, namely **Anthropocene** whose symptoms result from the scale of anthropogenic transformations of the environment (Stoner & Melathopoulos 2015).

---

140 Source: GDOS.gov.pl <https://ochronaprzyrody.gdos.gov.pl/files/artykuly/5450/WWD18_Infographic_PL_icon.pdf> [accessed: 5.09.2018].

3 The RETURN period

> *The mission now in architecture, as in all human endeavours, is to recover those fragile threads of connectedness with nature that have been lost for most of this century.*
>
> [James Wines 2008, p. 237]

The slogans of "*Returning to the River*", "*Regaining the River by the City*" or "*Urban Waterfront Renewal*", appeared in highly developed countries (the USA, Japan, Australia, Canada, Germany, Great Britain and others) already in the 1960s (Breen & Rigby 1994, 1996), and in Poland only at the turn of the 20th century. The pursuit was rather concentrated on reclaiming the impoverished post-industrial and post-harbour areas as well as on new forms of using river potential, especially its landscape and recreational advantages. The need for healthy life and effective leisure in clean environment turned into the social pressure for environment protection. Thus, the necessity of mitigation of environmental and flood risk as well as the interdependences between the catchment areas transformations and their consequences in the valleys were gradually comprehended.

The return period began during the "green breakthrough" of the 1970s and 1980s (in Europe) when we entered the stage of eco-hedonistic civilisation (Kosmala 2011, p. 5). The term accurately expresses the transformation of social values, although it was the hedonistic priorities of the development which induced raising the ecological awareness and its legislative implications.

We are at the beginning of the process. The conquest period has not finished yet but it is even growing stronger while the return to the principles of respect, coexistence and symbiosis is merely emerging. The prospect of population growth, anthropogenic climate catastrophy and water crisis requires undertaking decisive actions towards implementation of regenerative planning and circular economy.

3.1 Raising awareness

The processes of changes did not commence owing to declarations, directives, legal acts or regulations. The most important documents of global and European significance (treaties, declarations, conventions, directives, presented in Tab. 3.2) admittedly constitute "the milestones" but their creation and ratification followed a long way of raising social awareness: from realising the problems and social determination through the changes in values and development priorities to searching for the instruments of achieving the sustainable development goals (SDGs)[141]. The activators of the process were: science, education, and the media while the Internet invention proved to be the accelerator as the global network of information broadcast and thought sharing.

The stages and the range of changes concerning developing awareness, gradual legislative and spatial transformations comprise three priority aspects: **Red** – cultural and economic, **Green** – environmental and **Blue** – waterborne, which have been described in the following chapters: 3.2 Waterfronts revitalisation, 3.3 Environment regeneration, 3.4 Water governance. Despite their distinctiveness, they are inter-dependent and constitute one living space for people and the nature, therefore, they need to be perceived holistically. As Maciej Zalewski (2014, p. 10) says "It is of crucial importance to understand the complexity of interactions between the abiotic, biotic and socio-economic systems".

The modernist manifesto of CIAM (*Congrès International d'Architecture Moderne*) in the form of the Charter of Athens (1933) was a response to the deepening crisis of a traditional city. Noticing the crucial **leisure function** in cities as well as the postulate of healthy living space[142], drew the town planners' attention to the necessity of shaping the random spatial structures. At the beginning of the 20th century the aim of transformations was not to increase the space for greenery and water but first of all to enhance the access of air and sunshine to the built environment. As a result, the constructed housing estates provided much bigger areas of biologically active space than conventional city centre

---

141 The 17 Sustainable Development Goals and 169 targets were announced in: United Nations, 2015, *Transforming our World: The 2030 Agenda for Sustainable Development*, A/RES/70/1, <ttps://sustainabledevelopment.un.org/content/documents/21252030%20 Agenda%20for%20Sustainable%20Development%20web.pdf> [Accessed: 22.02.2020]

142 Le Corbusier – one of the founders of CIAM (*Congrès Internationaux d'Architecture Moderne – International Congress of Modern Architecture*) promoted the slogan of "the sun, space, greenery" – the three elements which by the mid-20th century had been considered as a luxury in European cities.

development (Januchta-Szostak 2015–2018). On the other hand, functionalism which dominated in the 20th-century town planning, downgraded rivers qualities to transport-technological functions[143]. Knowledge of the basic principles of hygrology (Byczkowski 1998) was negligible among urban planners, and the development of water engineering absolved them from responsibility for integration of hydro-graphic functions with city spatial planning.

Tab. 3.2: The profile of the most important international documents and events which influenced the change in city-river relation in terms of sustainable development as well as the enhancement of rivers and cities quality.

EUROPE	YEAR	THE WORLD
Athens Charter (*Charte d'Athènes*) CIAM, Athens, published in 1943	1933	
	1948	International Union for Conservation of Nature, IUCN
Founding the International Commission for the Protection of the Rhine (ICPR) in Basel	1950	
	1961	WWF (*World Wildlife Fund*)
European Water Charter	1968	
	1969	U Thant's report titled "*The Problems of Human Environment*"- at the UN session
	1971	*Ramsar Convention on Wetlands* concerning the wetlands of international significance, especially as the waterfowl habitat
	1972	*The Declaration of the United Nations Conference on the Human Environment*
HELCOM – the Convention on the Protection of the Marine Environment of the Baltic Sea Area (Helsinki Convention)	1974	

(*continued on next page*)

---

143 Only the improvement in water quality at the end of the 20th century made it possible to regain river valleys for the inhabitants' leisure needs.

Tab. 3.2: Continued

EUROPE	YEAR	THE WORLD
	1976	UN-HABITAT I (*United Nations Human Settlements Programme*) – the UN agenda concerning the urbanisation and human settlements, sustainable settlement program, Vancouver (*Vancouver Declaration on Human Settlements, the Habitat Agenda*)
Directive 79/409/ the EEC on the protection of wildfowl	1979	
ICP *Waters*, The International Cooperative Programme for assessment and monitoring of the effects of air pollution on rivers and lakes	1985	
Directive EIA 85/337 the EEC on the assessment of the effects of certain public and private projects on the environment	1987	"*Our Common Future*" the report by the Brundtland Commission
Foundation of *European Environment Agency* (*EEA*)	1990	IPCC – (*International Panel on Climate Change*) published a report including distressing results of research on global warming process and indicating human economic activity as the main cause of the changes
Green Paper on the Urban Environment		
Directive 91/271/the EEC on municipal sewage treatment	1991	
Directive 91/676/the EEC concerning water protection against the nitrate contamination from agricultural sources (Nitrates Directive)		
The UN Convention on the Law of the Non-Navigational Uses of International Watercourses or UN Watercourses Convention – on the protection and use of the cross border watercourses and international lakes	1992	Earth Summit in Rio de Janeiro – the UN Conference on Environment and Development (*Rio Declaration on Environment and Development*). Agenda 21 – *Global Programme of Action on Sustainable Development*; United Nations Framework Convention on Climate Change – UNFCCC
The Convention for the Protection of the Marine Environment of the North-East Atlantic		
Directive 92/43/EEC on the protection of natural habitats and wildlife (*Habitats Directive*). Program Natura 2000		

Tab. 3.2: Continued

EUROPE	YEAR	THE WORLD
	1993	The UN *Convention on Biological Diversity* – CBD
The Charter of European Sustainable Cities and Towns Towards Sustainability - Aalborg Charter	1994	
	1996	UN-Habitat II – elaborating on the principles of sustainable development of human settlement. (*Istanbul Declaration on Human Settlements*)
	1997	The implementation programme of Agenda 21
		UNESCO Declaration on the Responsibilities of the Present Generations towards Future Generations
Convention on Access to Information, Public Participation in Decision-making Processes and Access to Justice in Environmental Matters	1998	
The New Charter of Athens, the European Council of Town Planners, Athens		
European Landscape Convention Florence	2000	*Millennium Development Goals*, MDG) – eight goals established within the UN Millennium Project
Directive 2000/60/EC on establishing the framework for the Community action in the field of water policy (Water Framework Directive)		
Directive 2001/42/WE on the assessment of the effect of some projects and plans on the environment (Strategic *Directive EIA*)	2001	
	2002	The World Summit on Sustainable Development (WSSD) in Johannesburg - the UN Conference, Declaration on sustainable development, implementation plan of *Millennium Development Goals*
New Charter of Athens: (update) the European Council of Town Planners' Vision for Cities in the 21st century, Lisbon	2003	
Directive 2004/35/EC on Environmental Liability	2004	

(*continued on next page*)

Tab. 3.2:   Continued

EUROPE	YEAR	THE WORLD
	2005	The assessment of the ecosystem change for human well-being: *Millennium Ecosystem Assessment* (MEA). Ecosystem services
Leipzig Charter on Sustainable European Cities	2007	*The Baltimore Charter for Sustainable Water Systems* also concerning sustainable water management in American cities
Directive 2007/60/ EC on the assessment and management of flood risk (Floods Directive)		
Directive 2009/147/EC on the protection of wildfowl (The Birds Directive)	2009	
OECD (*Green growth and sustainable development*)		
Directive 2011/92/EU on the assessment of certain public and private projects on the environment	2011	
OECD (*Green Growth Strategy*)	2012	The UN Conference on Sustainable Development Rio+20 "*The future we want*"
The EU strategy of adaptation to climate change	2013	
OECD (*Principles on Water Governance*)	2015	The UN Sustainable Development Summit in New York. Agenda for sustainable development 2030 "*Transforming our world*", seventeen Goals of Sustainable Development for the years of 2015–2030.
	2016	UN-Habitat III - The New Urban Agenda (NUA), Quito Declaration on Sustainable Cities and Human Settlements for All and Implementation Plan. Conference in Istanbul. The Principles for Water-Wise Cities, International Water Association
	2018	Beginning works on "The World Pact for the Environment"

The significance of water was appreciated again in the second half of the 20th century in the **sustainable development** context. The UN programme concerning human settlement in the subsequent agendas (UN-Habitat I – 1976, UN-Habitat II – 1996 and UN-Habitat III – 2016) distinctly emphasised the importance of the **environment quality for the economic growth and life comfort in cities**. The postulates of CIAM were modified by the European Council of Town Planners (later: European Council of Spatial Planners) in the New Charter of Athens[144] (1998, 2003), and amended by the Leipzig Charter on Sustainable European Cities (2007). Water and greenery began to become key determinants of urban environment quality and their contamination has been perceived as a barrier to city development (Green Paper 1990). The role of the public opinion has grown along with social participation in public space and environment governance (Directive 2003/35/EC).

Environmental aspects of town planning, already mentioned in 1969 by Ian McHarg[145] in his book titled *Design with Nature*, for a long time remained in the sphere of scientific postulates which were difficult to be implemented in Europe. The same year, **Sithu U Thant's Report titled** *The Problems of Human Environment* reverberated when presented at the UN General Assembly[146], raising awareness of global environmental threats, activating pro-ecological movements[147] and initiating **"Green Breakthrough"** decade[148] in the United

---

144 One of the basic instructions of the Charter on shaping the urban environment was to treat a city as an ecological system where the non-renewable resources extraction and waste production should be limited. The New Charter of Athens (2003) mostly emphasises identity, connectedness and attractiveness of urban landscape (Januchta-Szostak 2011b).
145 Ian McHarg (1969) featured ecological aspect of urban space, providing foundations for methodology of environmental impact assessment and Green Urbanism concept, later developed by Timothy Beatley (*Green Urbanism*, Beatley 1997, 2000).
146 The report titled *The Problems of Human Environment* (resolution 2398) was presented by the UN Secretary-General, Sithu U Thant, at the General Assembly session on 26th May 1969. For the first time in history, the document made the public opinion aware of the data revealing the destruction of the natural environment and its detrimental consequences; it appealed to all countries for sustainable use of the Earth resources and for efforts to protect ecosystems.
147 *International Union for Conservation of Nature* (IUCN) was founded already in 1948 and WWF (*World Wildlife Fund*) in 1961.
148 It was initiated by signing the National Environmental Policy Act (NEPA) by the president Richard Nixon on 1st January 1970, including procedural requirements for preparation for the environmental assessment (EAS) and environmental impact statement (EISS). The same year, *Environmental Protection Agency (EPA)* was founded along with the Council on Environmental Quality. The detailed regulations and environment protection plans are created in particular states and approved by EPA (Januchta-Szostak

States. That led to formulating i.a. **Ramsar Convention on Wetlands** protection (1971) and **Stockholm Declaration** (UN 1972) on humans' natural environment, indicating responsibility of the humankind for the environment as well as the necessity of undertaking actions to protect and permanently monitor its condition in the whole world[149]. In 1987 The Brundtland Commission[150] developed a report *Our Common Future*, which contributed to convening the **Rio de Janeiro Earth Summit in 1992**[151] (UN 1992). **Agenda 21** was approved at the summit as a programme on protection and management of natural resources including i.a. the tasks concerning sensitive ecosystems management and protection of inland waters quality.

In Europe, however, by 1980s great hydro-technical undertakings had been performed without any consideration for the environmental impact. Only the directive EIA 85/337/EEC in 1985 obliged the European countries to assess the effects of certain public and private projects on the environment as well as the impact of certain plans and programmes (Directive 2001/42/EC). In the following years, a few key EU directives contributed to the environment protection, including primarily: directive 91/271/EEC concerning urban waste treatment (1991), the Nitrates Directive 91/676/EEC (1991), the Habitats Directive 92/43/EEC (1992) and the Birds Directive 2009/147/EC[152], constituting the basis for **Natura 2000 programme** (2009), or Directive 2004/35/EC on Environmental Liability (2004). Rising ecological awareness and intensification of international cooperation on environmental protection[153], as well as developing and using

---

211b, p. 53). The *European Environment Agency* (EEA) was established only 20 years later, in 1990.

149 The UN Stockholm Conference resulted in *United Nations Environment Programme* (UNEP) – UN agenda established by UN General Assembly resolution 2997 issued 16th December 1972.

150 The World Commission on Environment and Development, founded in 1983 by Gro Harlem Brundtland, by invitation of the UN Secretary General.

151 Three other key documents were enacted at the Earth Summit in 1992: UN Framework Convention on Climate Change (UNFCCC), Convention on Biological Diversity (CBD) and the *Rio Forest Principles* (the Non-Legally Binding Authoritative Statement of Principles for a Global Consensus on the Management, Conservation and Sustainable Development of All Types of Forests, 1992).

152 The first version of the directive was created 30 years earlier in 1979. The directives EEC 79/409/EEC issued on 2nd April 1979 (*Directive on the Conservation of Wild Birds*).

153 The analysis of the agreements from the mid-19th to the beg. of the 20th century was performed by Ronald B. Mitchel (2017) within the project titled *International*

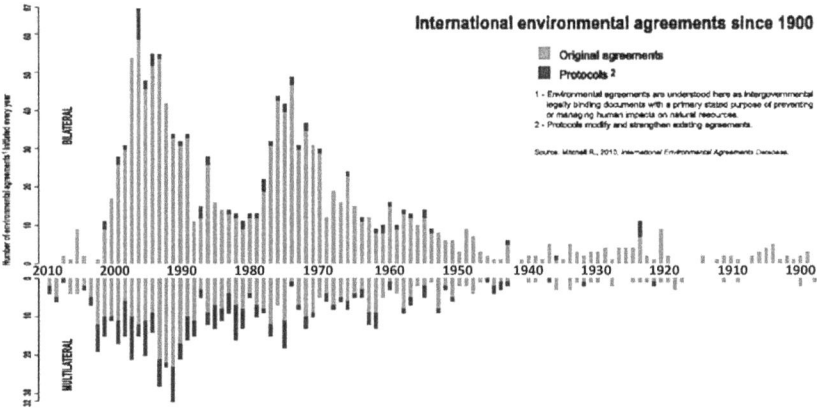

Fig. 3.21: The number of bi- and multilateral environment agreements in Europe since the beg. of the 20th century, developed by A. Januchta-Szostak based on: EEA, 2010, *Environmental Agreements since 1900*, source: <https://www.eea.europa.eu/data-and-maps/figures/environmental-agreementssince-1900> [accessed: 5.08.2018]

rivers, have manifested in considerable increase in the number of multilateral environmental agreements, especially since the 1990s. The following diagram (Fig. 3.21) clearly illustrates two peaks of "green breakthroughs" in the years 1970–1980 and 1990–2000.

Consecutive Earth Summits organised by the UN in Johannesburg (UN 2002) and in Rio de Janeiro (UN 2012) popularised ideas and particularised the directions of **sustainable development** on the basis of research development and new challenges of the 21st century. **Climate change tendencies diagnosed in IPCC reports**[154] were also reflected in the 2030 Agenda for Sustainable Development, titled *Transforming our World* (United Nations 2015), approved at the UN Sustainable Development Summit in New York in 2015.

One of the most threatening water risks, caused by the **global climate change**, is the increase in extreme weather phenomena, including torrential rains, heat waves, droughts, glaciers melting and rising level of seas and oceans. The EU

Environmental Agreements Database. Project (Version 2017.1), Eugene, OR: University of Oregon <https://iea.uoregon.edu/> [accessed: 3.08.2018].

154 In 1990, Intergovernmental Panel on Climate Change (IPCC) published the first report, including alarming results of the research on global warming and indicating human economic activity as the main cause of the changes.

strategy on adaptation to climate change (EC 2013) especially emphasises preparations for extreme weather phenomena as well as reductions in social and economic cost which they entail. It also emphasises the necessity of considering the adaptation issue in town planning and development as well as in programmes on preventing urban floods (Januchta-Szostak 2016). The research on adaptation solutions, conducted by EEA (EEA 2016b), revealed that water governance is the priority sector for many countries, while multi-functionality of solutions enables such a use of blue-green urban infrastructure that reduces water problems.

Waterborne problems, concerning both quality (rivers condition, kinds of contaminants), as well as quantity (droughts, floods, water sources governance) were neglected in terms of legal regulations of the EU for the longest time[155]. Only at the beginning of the 21st century two crucial directives were enforced: the directive 2000/60/EC of the European Parliament and the Council (2000) on establishing the framework for the Union water policy (known as The **Water Framework Directive**) as well as the directive 2007/60/EC of the European Parliament (2007) on the assessment and management of flood risks (known as **Floods Directive**). The Water Framework Directive (2000) not only normalised the frame of water economy and policy but also obliged the EU member countries to monitor and enhance water condition[156]. While the Floods Directive (2007) dramatically changed the paradigm of flood protection, which had prevailed since the 18th century, now allowing rivers to overflow and respecting the need to ensure and even expand floodplains. Its aim is to reduce and manage the flood risks and minimise negative flood consequences in the EU countries.

The development of **ecology and environmental ethics** was a vital stimulator of the changes. Understanding the ecosystems dynamics (Holling 1978, Peterson et al 1998) drew the researchers' attention to the possibilities of following the natural cycles of the natural world in cities development (Tjallingii 1995). The advantages of the natural environment, which had been lost due to its degradation, were noticed and economically appreciated as **ecosystem services** rendered

---

155 In the USA in 1972, *Clean Water Act* 1972 was enacted, establishing ambitious goals of regenerating and keeping the chemical, physical and biological water purity, eliminating contamination sources and achieving good water condition which would allow for fishing industry, healthy wildlife as well as water leisure opportunities (Otto et al 2004). In Europe, similar goals were established in Water Framework Directive (2000) which began to be in force almost 30 years later, and in Poland it was transposed to Water Law Act and other acts only in 2011 (Januchta-Szostak 2017b).
156 The assessment of European waters condition performed in 2018 showed the enhancement in their quality but also the scale of degradation (EEA 2018).

Tab. 3.3: Environmental ethics and various approaches to the relation with the environment (developed by: Świątek 2017 based on Thompson 2000).

Anthropocentric approach	Egocentrism	We act to gain our own benefits.
	Homo-centrism	The biggest welfare for the biggest number of people. We care for the environment, aiming to gain benefits and pleasure for humans.
	Techno-centrism	Technology will solve all the environmental problems.
Non-anthropocentric approach	Physiocentrism	We protect the environment independently from human needs and preferences.
	Eco-centrism	We have responsibility of care for the whole environment, paying special attention to biosphere and ecosystems.
	Biocentrism	We care for all the members of the biotic community, as we are part of the nature.
	Theocentricism	We are part of spiritual development plan.

by urban natural capital (Daily 1997, Pedersen-Zari 2012). While the **development of ecological engineering** (Odum 1971) created the tools of designing sustainable systems, considering ecological essentials which integrate the needs of the society and the natural environment for mutual benefits (Bergen et al 2001).

The anthropocentric perspective of the environment perception, dominating in the conquest period, has been replaced with alternative physio-, eco- or biocentrism (Tab. 3.3). The awareness of the fact that humans are not the owners of the Earth but belong to sophisticated ecological systems[157] and the deteriorating environment conditions result from disturbing the ecosystems balance[158], due to human activity, initiated formulating the rules of "regenerative development"[159].

The perception of water based on considering its advantages and risks is still the prevailing approach while the noticeable return to respect for the

---

157 The books by Eugene P. Odum: *Fundamentals of Ecology* (1953) and by his brother Howard Odum: *Environment, Power, and Society* (1971) had fundamental significance for the development of ecology.
158 The research titled *Millennium Ecosystem Assessment* has been the biggest undertaking which aims to identify the goods and services of the environment (published in: Hassan et al 2005; Carpenter et al 2005; Chopra et al 2005).
159 In the 1970s, the idea of regenerative design was developed by T. Lyle (1994), the author of i.a.: *Regenerative Design for Sustainable Development*; on the basis of his works, Centre for Regenerative Studies was founded (Świątek 2017, p. 10).

environment does not mean "sanctity of the nature" but is rather based **on economic grounds**. *Homo œconomicus*[160] requires rational arguments to cease to exploit the environment and bear the costs of its restoration. The necessary natural life-support processes have been therefore evaluated as "ecosystems services"[161] and expressed in economic jargon as the contribution of natural ecosystems to widely understood human well-being or as a profit generated by the natural assets (Norgaard 2010). In the economic-technical terms, the greenery structures became "Green Infrastructure" while the "ecological engineering" has been defined as the use of the natural environmental processes for the purpose of gaining benefits both for the humans and the environment itself (Odum 1971, p. 331). The integration of the environment with the economy was distinctly expressed by OECD in Declaration on Green Growth (OECD 2009) and in *Green Growth Strategy* (OECD 2011) orientated towards achieving economic growth and development while ensuring that natural assets continue to provide the resources and environmental services which enhance human life quality. At the Sustainable Development Summit Rio+20 Member States decided to launch a process to develop a set of Sustainable Development Goals (SDGs) build upon the Millennium Development Goals. The Conference also adopted ground-breaking guidelines on green economy policies.

However, as Richard B. Norgaard (2010) emphasises, "the metaphor of nature as a stock that provides a flow of services is insufficient for the difficulties we are or in the task ahead". Increasing **ecological footprint** (Wackernagel and Rees 1996)[162] of cities and entire humankind has already considerably exceeded ecosystems productivity[163] as well as the environment ability to regenerate.

The raised awareness, underpinned by the requirements of the EU directives on the assessment of the effects on the environment, protection of biodiversity, enhancement of surface waters quality and providing space for water, caused the change in priorities of the development of river valleys in the city. The urban

---

160 The paradigm of a rational human being (*homo œconomicus*) was formulated by John Stuart Mill in the 19th century and adopted by most of economic theories.
161 Costanza et al (1997) calculated that the total value of the services of the environment exceeds the value of the GWP by at least 80 %.
162 An article by William Rees in 1992, tilted *Ecological Footprints and Appropriated Carrying Capacity: What Urban Economics Leaves out* was the first scientific publication on ecological footprint.
163 In 2014 ecological footprint of humankind was 69.6 % higher than biological efficiency of the Earth (Global Footprint Network 2018).

waterfront transformations are mainly orientated towards (Januchta-Szostak 2017c):

- **waterfront revitalisation**, connected with the enhancement of **life quality and cultural environment** (Red), including socio-economic regeneration (through leisure-tourist activation and commercial use of water vicinity) and landscape regeneration (redevelopment of embankments and banks, creating attractive and accessible public spaces);
- **natural regeneration of the river valleys** (Green), orientated towards recovery of the quality and biodiversity of water and waterside ecosystems as well as continuity of valley green corridors in connection with ecological education and restoration of landscape values;
- **water quality and flood risk management** (Blue) with the use of blue-green infrastructure, enabling restoration of the natural hydrological cycle in urban catchment areas (retention, infiltration, water purification and evaporation).

The contemporary approach to the development of river valleys in cities features multi-functionality of solutions aiming to achieve all the above mentioned purposes, though reaching the eco-hydrological objectives requires actions considerably exceeding the urban sections of river valley, namely **restoration of the connections between the watercourses and their catchment areas**. In order to regenerate urban rivers and their coexistence with the city, it is not only crucial that the character of the particular sections of the valley should be determined in terms of socio-economy, culture, nature and hydro-morphology, but first of all, it is vital that the **planning and spatial development should be integrated with water governance in the entire catchment areas**.

3.2 Waterfronts revitalisation[164]

Revitalisation (*urban renewal*) refers to the processes of "invigorating" urban structures being in crisis due to the multiannual neglect which led to loss of their social and economic vitality and landscape appeal. Polish legal acts[165] do

---

164 The author discussed the issues in i.a.: *Poznań Waterfront – Warta Valley. Revitalisation of the Relationship with the River* (2011a); *Woda w miejskiej przestrzeni publicznej...* (2011g); *Frontem do rzeki. Współczesne tendencje w zagospodarowaniu frontów wodnych i dolin rzecznych w miastach* (2014a); *Regeneracja dolin rzecznych w miastach* (2017c).
165 "revitalisation is a process of bringing degraded areas out of crisis, which is performed in a comprehensive way by integrated actions for the sake of the local community, space and economy, territorially concentrated, conducted by revitalisation stakeholders on

not mention revitalisation of rivers, however, revitalisation programmes often concern riparian districts and post-industrial terrains.

In the 1950–60s, **the pace of industrialisation** of riparian areas **decreased** sharply. Technological progress, development of car and air transport reduced the importance of navigation leading to the abandonment of the river banks by industrial plants which left behind degraded development and environment. The decline of waterfront areas was accelerated by transferring navigation and shipping functions from the centres to the suburbs[166], e.g. from London centre to Tilbury and Felixstowe, from Rotterdam to Botlek and Europoort, from Amsterdam to Haarlem etc. Within 20 years (1960–80) the majority of harbours in big European cities emptied (Lorens 2001) while the zones of degraded post-industrial brownfields, blocking access to the rivers, required expensive transformations[167]. Consequently, the deteriorated waterfronts, merely accessible and lacking connection with the urban tissue, lost their visual and functional attractiveness. In a broader perspective, the functional, landscape, economic and social city relations with the river were severely affected (Januchta-Szostak 2011a).

Regaining the waterfronts of European cities

The waterfront terrains became a testing ground for revitalisation processes for the following 50 years. **Reclaiming the space** of the former harbours, abandoned warehouses and factories was initiated by the USA with the project called *Baltimore's Inner Harbor* (1963) which became a catalyst for urban waterfronts development: Boston, Toronto (Harbour front, 1972), New York (Battery Park City, 1979, Hudson and East River banks in Manhattan), Vancouver (Grandville Island, 1979) and many others (Wrenn et al. 1983, Breen and Rigby 1996, Hudson 1996, Lorens 2001, 2007). The socio-economic changes and the developmental priorities shift furthered restoration and renewal of the riparian districts. Production functions have been replaced by housing and service development and sometimes, communication infrastructure[168]. Also, the increase in

---

      the basis of revitalisation programme of the commune" (Legal act 9th October 2015 on revitalisation). The criterion for determining the degraded area is concentration of negative social phenomena (Art. 9. 1).

166 In the 1960–70s a container ship system was implemented but city harbours were too shallow and too narrow for huge container ships.

167 Between 1960 and 1980, all the London docks were closed while their zone of 21 km² (!) separated London from the Thames.

168 The riparian zones were used for location of expressways in the city centres, e.g. in Barcelona, Warsaw or Szczecin.

the amount of free time eventuated in growing need for leisure spaces. In densely built-up city centres, reclaiming waterfronts gave an opportunity for opening a vast river view and development of functions which interrelate with waterside leisure.

The European examples of pioneer revitalisations of harbours, initiated in the 1980s, are i.a. London *Docklands* by the Thames (since 1981) and *Kop van Zuid* by the Meuse in Rotterdam (since 1987)[169]. Big investments in London and Rotterdam, mostly focusing on the economic and infrastructural factors, "showed the way" of coordination and financing to achieve their main purpose: the transformation of city image, attracting both investment (London – cf. Chap. 4.3) and new inhabitants (Rotterdam – cf. Chap. 4.2).

Revitalisation processes of waterfronts in Europe and in the world have been analysed and described by many researchers, who emphasise inevitability of urban "recycling" (Meyer 1999; Marshall R. 2001; Bruttomesso 1999, 2011) as well as spectacular effects of urban waterfronts renewal (Bruttomesso 1993; Breen and Rigby 1996; Kochanowski 1998, Fisher and Benson 2004); but they also mention social problems resulting from gentrification of districts (Jadach-Sepioło 2007) along with consequences for the landscape e.g. "Manhattanisation" effect (Ibelings 2003). The key success factors were described in a document titled *10 Principles for Sustainable Urban Waterfront Development*)[170].

At the turn of the 20th century, the areas of the former downtown river and sea harbours underwent metamorphoses, e.g. Vell Harbour in Barcelona (1992), Porto Antico in Genoa (1992), Københavns Havn and Inderhavnen in Copenhagen, Aker Brygge and Tjuvholmen in Oslo (Fig. 3.22, 3.23), the banks of the IJ River and IJmeer Lake (formerly the sea) in Amsterdam (since the end of the 1990s, Fig. 3.24, 3.25), or HafenCity by the Elbe in Hamburg (since 2004). The cities had reached such a developmental stage that regaining the degraded riparian terrains became not only profitable but even necessary (Januchta-Szostak 2017c).

---

169 The processes of waterfront revitalisation of Rotterdam and London were more comprehensively described in Chap. 4 "Responsible cities – vital rivers".
170 *10 Principles for Sustainable Urban Waterfront Development* was approved at *Urban 21* international conference under the aegis of the UN in June 2000 and accepted at Waterfront Expo in Lisbon in 2007.

Fig. 3.22: Astrup Fearney Museet at Aker Brygge in Oslo (photo by A. Januchta-Szostak)

Fig. 3.23: Former harbour Aker Brygge waterfront in Oslo (photo by A. Januchta-Szostak)

Fig. 3.24: New development of Zeeburg district in Amsterdam - Jawa (photo by A. Januchta-Szostak)

Fig. 3.25: "NEMO", New Metropolis in the Eastern Dock refers to harbour past of Amsterdam (photo by A. Januchta-Szostak)

Fig. 3.26: Park of the Nations by the Tagus in Lisbon with hardly accessible bank line (photo by A. Januchta-Szostak)

Revitalisation of the southern bank of the **IJ River in Amsterdam**, initiated at the end of the 1990s by the redevelopment of the **Eastern Docks**[171] (*Oostelijk Havengebiet*), presents both the Dutch panache approach to the space "recycling" and the capability to create an identity of a place. The Zeeburg district was settled on four post-industrial islands/peninsulas (KNSM, Borneo, Java and Sporenburg), transformed into appealing housing complexes for 17,000 inhabitants. Greenery shortage[172] was compensated by the open water space, covering two-thirds of the new district area (Januchta-Szostak 2009a). The commercial and image success of the investment inclined the city authorities to further expand on the IJmeer Lake

---

171 The Eastern Docks in Amsterdam were constructed in the 19th century following the development of goods trade and transport from Eastern India as well as the new location of the railway station.
172 The surface area: 69 ha, including green areas: 9.3 ha (13.40 %), public space: 30.29 ha (43.9 %), housing development: 21.2 ha (30.70 %), services: 8.21 ha (12.01 %).

waters, which resulted in constructing IJburg (construction started in 1996) – a district planned for 45,000 inhabitants, with a diverse functional structure, located on six artificially created islands. Satisfying the housing needs within the metropolis was an unquestionable advantage of both investments. Keeping the cosy scale of the development, which features the character of old-time Amsterdam, is undeniably another positive aspect of the architecture of Zeeburg (Fig. 3.24) and IJburg. The district construction plans provoked controversies though because of the scale of interference into water environment, however, for the majority of the Amsterdam's public they were not sufficiently considerable to make them participate in referendum[173]. Whereas abandoning the large Markerwaard construction, included in the development plan of Zuiderzee Bay by Cornelis Lely (1891), was a sign of respect for ecological priorities. The extent of urbanisation plans was gradually limited and finally abandoned by the Dutch authorities. In 2016 the previously planned polder was replaced by implementation of Marker Wadden project, aiming to create an archipelago of artificial islands and wetlands serving ecosystems strengthening and biodiversity enhancement in Markermeer reservoir. The islands are supposed to be a tourist-accessible birds sanctuary.

Also in the 1990s, the investment in **Bilbao**[174] **by the Nervion** river, indicated another direction for the revitalisation of riparian areas. The local authorities' idea was to use modern spectacular architectural works in order to promote the city image and tourist attractiveness of the region[175]. The construction of groundbreaking significance was the **Guggenheim Museum** designed by star architect Frank Gehry, built in 1994. The spectacular, even shocking facility, displayed owing

---

173 The opponents of the IJburg district construction pointed out the negative consequences of the investment for the IJ lake environment. The results of the referendum conducted in 1997 were negative (the majority voted against the construction), but due to the insufficient number of the participants, the authorities upheld the decision to build the district. <https://www.iamsterdam.com/en/living/about-living-in-amsterdam/neighbourhoods/ijburg> [accessed: 9.08.2018].

174 The collapse of the metallurgical and shipyard industry led to unemployment increase by 25 % and economic break down. The reaction to the crisis consisted in functional and image transformations which changed the former industrial centre into the cultural centre of the region. In Bilbao, the emerging architectonic facilities, designed by celebrities of modern architecture, include spectacular underground stations (project by Norman Foster) and the footbridge (by Santiago Calatrava). More:<https://archirama.muratorplus.pl/encyklopedia-architektury/efekt-bilbao,62_3345.html> [accessed: 9.08.2018].

175 Similar projects were realised much earlier by Sydney authorities in Australia, constructing iconic opera edifice in modern expressionism style on the Bennelog Point from 1957 to 1973 (project by Jørn Utzon & OveArup).

Fig. 3.27: The European Solidarity Centre (project by FORT) - the icon of the Young City in Gdańsk (photo by A. Januchta-Szostak)

to the river foreground, enlivened the tourist traffic and contributed to creating new vacancies in the service sector. That cultural phenomenon called the **Bilbao effect** became a pattern which many cities tried to follow while riparian areas developed into the places of **architectural icons display as well as appealing public spaces**. In Amsterdam i.a. the facility called "Nemo"– (New Metropolis Museum designed by Renzo Piano, 1997) was constructed and shaped like a ship "anchored" in the Eastern Dock (Fig. 3.25). In Hamburg, the edifice of the Philharmonic Hall (project by Herzog & de Meuron) is the showcase of HafenCity while the European Solidarity Centre (project by FORT) is the icon of the Young City in Gdańsk (Fig. 3.27). Architectural icons of many cities clearly refer to their harbour past. The dialogue with water, which conveys universal message, is a material of creating cities identity to the same extent as uniqueness of their architecture, landscape and urban structure layout, history and social diversity (Januchta-Szostak 2009).

Another factor furthering the waterfront transformations, especially building impressive amenities and architectural complexes, resulted from the necessity of preparation for great cultural events (Januchta-Szostak and Biedermann

2014) such as Olympic Games (e.g. in Barcelona in 1992, or in London in 2012), world exhibitions (Expo'1992 in Seville by the Guadalquivir River, Expo'1998 in Lisbon[176] by the Tagus (Fig. 3.26), Expo'2008 in Saragossa by the Ebro River) or the title of European Capital of Culture (e.g. Rotterdam – 2001; Copenhagen – 1996, Wrocław – 2016). Admittedly, **Copenhagen** is a sea harbour but its urban waterfront revitalisation is worth mentioning here not only because of the scale of the transformations (Øresund region) but also the innovative pro-social approach, initiated by Jan Gehl[177] (Gehl 1987) as well as the sensitivity to ecological aspects. The revitalisation works, commenced in the 1980s, transformed the harbour docks of Holmen[178] into a new cultural centre[179]. The architectural icons were erected by the water: the National Film Centre, the new Architecture Centre, the National Library (project by Schmidt Hammer Lassen Architects) and the Opera (project by H. Larsen). The city metamorphosis features human scale and high quality of public space.

Waterfronts revitalisation in Poland

In Poland, the revitalisation processes of urban riparian spaces gained new momentum only in the 21st century and the vast majority of projects are focused on the socio-economic priorities[180]. **Gdańsk** can be proud of the transformation

---

176 In Lisbon, the revitalisation works covered the entire post-industrial terrain by the Tagus riverbank where the Nations Park and Expo City were created.
177 The first research and revitalisation proposals of the waterfront in Copenhagen in the 1970s concerned the historical centre, located by the Nyhavn, the main harbour canal where the laboratory of creating the public space was founded under the supervision of Jan Gehl and in the mid-1980s the effects of new goals achievements were already visible: regaining the identity of the place and creating a tourist attraction (Petryshyn 2015).
178 Just as in Amsterdam, the harbour area of Holmen, were created in the process of transformation and reclamation of shallow waters near the shore.
179 In 1996, Copenhagen was the European Capital of Culture, which contributed to the development of cultural institutions, tourist traffic intensification and creating the new city image.
180 It has been proved by the results of the research on the revitalisation processes of urban riparian spaces in northern Poland, conducted by the author within the REURIS project (2009–2011), published in i.a.: REURIS, 2012, Urban Rivers – Vital Spaces. Manual for urban river revitalisation – implementation, participation, benefits. 2012, Bydgoszcz, pp. 327; source: <https://www.yumpu.com/en/document/view/25724375/manual-for-urban-river-revitalisation-central-europe> [accessed: 4.03.2019].

Fig. 3.28: The revitalisation of the Motława River waterfront in Gdańsk - Granary Island (photo by A. Januchta-Szostak)

of Wyspa Spichrzów (Granary Island) and Ołowianka[181] (Fig. 3.28) by the Motława as well as the wide-reaching revitalisation programme in the former Gdańsk shipyard area which since 2011 has been transformed into the Young City district[182] (Fig. 3.27). In **Szczecin,** since the 1990s the revitalisation works of Łasztownia, Wyspa Zielona and Kępa Parnicka have been in process. In **Warsaw,** revitalisation includes i.a. Czerniakowski Harbour and Promontory as well as the boulevards along the Vistula river. Revitalisation of Nadodrze in **Wrocław**[183] (Fig. 3.29) also produced spectacular effects which have been intensified owing to the title of the European Capital of Culture, granted to Wrocław in 2016. The areas by the Odra river in **Opole** have undergone transformations as well. The Municipal Revitalisation Programme for **Kraków** included i.a. the areas by the Vistula river: Stare Miasto (Old Town) and Stare Pogórze-Zabłocie where the

---

181 In the refurbished facilities of the former power plant, The Polish Baltic Philharmonic was founded (project by KD Kozikowski Design, realisation 1998–2005).
182 The investment includes the transformation of the former post-harbour and post-industrial areas of 22 ha into a residential district, where there will also be shopping, service and entertainment space. More: Lorens 2007.
183 The pilot programme of Nadodrze revitalisation began already in 2004 and covered 110 ha.

Fig. 3.29: Xawery Dunikowski Waterfront in Wrocław (photo by A. Januchta-Szostak)

Vistula boulevards were created. In **Poznań**, the revitalisation effects of Śródka district by the Cybina canal, Chwaliszewo district as well as the extension of "Wartostrada" have been increasingly tangible.

The Programme of Revitalisation and Development of Bydgoszcz Water Junction[184] is an example of consistently realised "turning towards the river" strategy of Bydgoszcz city development, which has been implemented since 2016, activating the E-70 waterway and raising the waterfront attractiveness. The most known, realised example of comprehensive revitalisation of urban waterfront can be represented by the **Mill Island in Bydgoszcz**[185] (surface 6.5 ha, Fig. 3.30) previously the residence of Royal Mint and Mill industry and currently a green enclave in the heart of Bydgoszcz, surrounded by the Brda and Młynówka rivers as well as the picturesque development of Bydgoszcz Venice .

The purpose of the multi-stage revitalisation programme of Mill Island and its immediate surroundings was the restoration of its landscape and tourist

---

184 *Programme of Revitalisation and Development of Bydgoszcz Water Junction* was developed by MPU team in Bydgoszcz within the framework of *Exploiting Inland Waterways for Regional Development* project, financed by Interreg IIIB – Baltic Sea Region (In Water). The programme was the first such a document in Poland.

185 The project of Mill Island revitalisation was recognised in Poland and abroad, awarded the title of *Bryła Roku* in 2012, certificate in the 9th edition of the UN-HABITAT competition (United Nations Human Settlements Programme), prize in the "cooperation" category in EUROCITIES Awards 2011 and The World Leisure International Innovation Prize 2012.

Fig. 3.30: Mill Island in Bydgoszcz and reconstructed canal of Międzywodzie (photos by A. Januchta-Szostak)

potential[186]. As a result of the transformations, the historic buildings of mills, granaries and the mint were refurbished and adapted to new culture-making functions. Also, the island's accessibility was enhanced owing to the construction of three catwalks and the bridge renovation, along with the technical road, water supply-sewage system and hydro-technical infrastructure. The waterfronts of the Brda and Młynówka rivers were redeveloped, leaving promenades along President Narutowicz Waterfront while the old track of the Międzywodzie Canal was restored. The leisure amenities of the island have been enriched with an amphitheatre, a playground and, most importantly, with a new yacht-kayak marina, a hotel and a restaurant (Januchta-Szostak 2014a).

---

186 Objectives: stage 1: 2006–2007 – economic activation, stage 2: 2007–2009 – cultural heritage facilities renovation, stage 3: 2008–2010 – creating leisure infrastructure, stage 4: 2009–2011 – new marina construction. Based on the materials of Bydgoszcz Municipal Office: Revitalisation of Mill Island in Bydgoszcz for entrepreneurship development, sources: <www.mapa.funduszestrukturalne.gov.pl/projekt_pdf,id,112804><http://www.bydgoszcz.eu/miasto/aktualnosci/Drugi_etap_rewitalizacji_Wyspy_Mlynskiej.aspx> [accessed: 14.06.2014].

Fig. 3.31: The semi-natural, right Vistula riverbank in Warsaw – a valuable area of Natura 2000 in the centre of the capital - the world phenomenon and a unique space of contact with the nature for the inhabitants of the big agglomeration (photo by A. Januchta-Szostak)

River valleys in many Polish cities, unlike the highly transformed river sections in European agglomerations, retained a considerable potential of natural landscape. The **unregulated Vistula River in Warsaw**[187] (Fig. 3.31) is a global phenomenon. The left, densely urbanised Vistula bank displays cultural values of the capital. Whereas the right river bank, included in Natura 200 programme, is a unique space of contact with the nature for the inhabitants of the great agglomeration as well as the migration corridor (Fig. 3.32).

---

[187] The research on the management and use of the Vistula in Warsaw was conducted i.a. by M. Okołowicz (SGGW Warsaw), M. Wojnowska-Heciak (WA PW) and E. Maciejewska (WA PW) within the frame of their Ph.D. theses.

Fig. 3.32: Warsaw: the left (on the left) and the right (on the right) Vistula bank viewed from Poniatowski bridge (photo by A. Januchta-Szostak)

Fig. 3.33: Warsaw – boulevards by the Vistula. On the left: Kościuszkowski waterfront – a narrow bank zone separated from the city by a communication artery. On the right: the new section of the Vistula Boulevards – Powiśle (photo by A. Januchta-Szostak)

Polish cities have been compensating for the approx. 30-year backlog in redeveloping their relations with rivers by pursuing the revitalisation goals of western countries of the 1980–90s. New boulevards, harbours, marinas, culture and art facilities as well as waterside restaurants and cafes, enliven the social river space, referring to the cultural identity and contemporary users' needs. The high building intesity and greenery shortage are compensated by water vicinity and the appeal of the riverside landscape. The most crucial issues which enable social revitalisation of waterfronts are the **enhancement of their accessibility** for all the users and limiting vehicle traffic. Boulevards and pedestrian riparian promenades facilitate access to the river and relaxation by contact with water, while continuity of collision-free riparian routes may initiate the system of alternate pedestrian-cyclist communication. Equally important is to provide accessibility to the city from the waterways side as the unused harbour infrastructure has been largely degraded and leisure functions of waterfront require a suitable architectonic frame (Fig. 3.33). Both the high quality of waterside architecture and

Fig. 3.34: Cracow: the Vistula meander by Wawel – comparative profile by courtesy of R. Konieczny. On the left: a historical photo: Digital National Museum (1-G-4566-2). On the right: modern view (photo by R. Konieczny)

public spaces as well as comprehensive revitalisation of river valleys, contribute to the enhancement of the city image, tourist attractiveness and its inhabitants' life quality (Schneider-Skalska 1997; Januchta-Szostak 2011b).

Regardless of the place and scale, all the examples of waterfront revitalisation mentioned above focused on cultural and socio-economic goals: **reclamation of the waterside terrains for new commercial investments, restoration of urban riparian structures as well as economic and tourist activation.**

Great harbours, like Rotterdam or London, were especially affected by the loss of **identity and liveliness of shipping**. Therefore, revitalisation investments tended to search for a **new image** while the nostalgia feeling about the lost economic vitality of the rivers facilitated activation of new leisure and tourist functions of navigation (Fig. 3.34). Transformation of waterfronts, especially within big cities, is a long-lasting and costly process, often limited by various factors. The slow pace of changes sometimes brings unexpected advantages of backwardness, like in the case of Warsaw which benefited from the shift in the paradigm of shaping river valleys, preserving the invaluable assets of wild nature on the right Vistula bank.

3.3 Environment regeneration

The **Convention on Wetlands** (Ramsar treaty 1971), signed in Ramsar, Iran in 1971, is the only international environment treaty devoted to a particular ecosystem – wetlands[188]. The Convention is significant as it concerns the issues of

---

188 The term wetlands, or freshwater-swamp areas, in accordance with the Convention are: *"areas of marsh, fen, peat-land or water, whether natural or artificial, permanent or*

the level of wetlands degradation at a global scale[189] and the value of the ecosystem for the environment, the mankind and the economy. The assessment of the services of global ecosystems and natural assets, conducted by R. Costanza et.al. (1997), revealed that the two types of ecosystems significantly decreased the value of rendered services: freshwater-swamp areas and river estuaries due to the degradation of watercourses, valleys and catchment areas. The restoration of the natural environment of rivers and wetlands is becoming an essential element of national and international strategies, included in water policy (Szałkiewicz et al 2018). Especially, given the fact that the regenerated and well managed river systems can mitigate the negative consequences of ground use and extreme hydro-meteorological phenomena as well as contribute to biodiversity rise also in cities[190].

Natural regeneration[191] **of the environment of river valleys** as the process of recovering welfare should be considered in eco-hydrological aspects (rehabilitation, re-naturalisation, restoration of indigenous species) as well as cultural and socio-economic (revitalisation, revalorisation). "'River restoration' describes a wide range of activities aimed at restoring the natural state and functioning of rivers and the water environment" (*Rivers by Design* 2013, p. 4)[192]. **Small-scale** site specific projects include replacing the concrete embankments with natural materials and plants, liquidation of invasive species and restoring the habitats of local wildlife. **Medium scale** projects aim to improve river continuity and connectivity with floodplains and natural flood risk management. They involve

---

temporary, with water that is static or flowing, fresh, brackish or salt, including areas of marine water the depth of which at low tide does not exceed six metres". The Convention was signed by 90 % of the UN member states. Poland ratified the Convention in 1978. Source: GDOS 2013-12-23 <https://www.gdos.gov.pl/konwencja-ramsarska> [accessed: 1.08.2018].

189 Only from 2009 to 2016, 33 % of water-swampland terrains disappeared in the world while in Europe as much as 45 % due to anthropopression (Hu et al 2017).

190 World Wetlands Day held in 2018 was themed as "Wetlands for Sustainable Urban Future". In 2016, the European Environment Agency published a report titled "*Rivers and Lakes in European Cities*" (EEA 2016a).

191 According to Merriam-Webster Dictionary regeneration means: 1 : an act or the process of regenerating; 2 : spiritual renewal or revival; or 3 : renewal or restoration of a body (https://www.merriam-webster.com/dictionary/regeneration) .

192 The European Centre for River Restoration is a pan European network of national river restoration centres and other members bound by a common mission to promote and enhance river restoration throughout Europe.

uncovering small urban watercourses, formerly canalised underground or changes in morphology of regulated riverbeds, recreating river meanders and restoring the natural erosion-sedimentation processes. **Big-scale** projects concern river corridors restoration and landscape planning, connecting river systems, recreating water-dependant ecosystems through hydro-morphological transformations of considerable valley sections and reclaiming flood plains (depoldering, dismantling or relocation of embankments). The purpose of the undertaken actions is restoring the variety of species and landscape forms of the local ecosystems and the environmental balance as well as recreating its retention capability, resilience and self-regeneration abilities. The actions aiming to recreate the river condition so that it is similar to the original natural state, which has been lost due to the watercourse engineering, is called **re-naturalisation** (Żelazo 2006, Żelazo and Popek 2002).

River regeneration at a regional scale

The results of centuries-old environmental transformations of big, engineered rivers such as the Rhine, the Thames, the Scheldt or the Meuse are difficult to reverse. The valleys of the rivers have lost the majority of invaluable ecosystems along with their self-purification capability while the dense settlement system has been the main source of contamination.

The European Environment Assessment, conducted in 2018, revealed that the water quality of the most highly urbanised European regions is still unsatisfactory. Comparison of water quality maps (Fig. 3.35) and the scale of watercourses transformations (Fig. 2.11) clearly shows the correlation between the location of areas with the highest level of river engineering and their water condition. "The core of the problem seems to lie in the fact that the considerable natural advantages of rivers and valleys were identified and appreciated a relatively short time ago, even by naturalists. Therefore, the requirements of nature protection were ignored in concepts and realisations of water-economic undertakings" (Żelazo 2006). Despite the intense reconstructive actions undertaken by Great Britain, Germany and Holland, rivers in the countries have still not regained their biological balance.

The necessity of water quality enhancement and re-naturalisation of river valleys, resulting from the recommendations of the Water Framework Directive (WFD 2000), became the foundation for actions undertaken with the aim to regain the natural-like river condition in many European countries at a regional, national and international scale. Implementation of environmental directives and WFD caused evident increase in the number of realised projects on river regeneration in Europe after 2000 (Szałkiewicz et al 2018, Fig. 3.36).They often may be still fragmentary actions, which result from local initiatives rather than sustainable strategies of rivers revitalisation and regeneration[193].

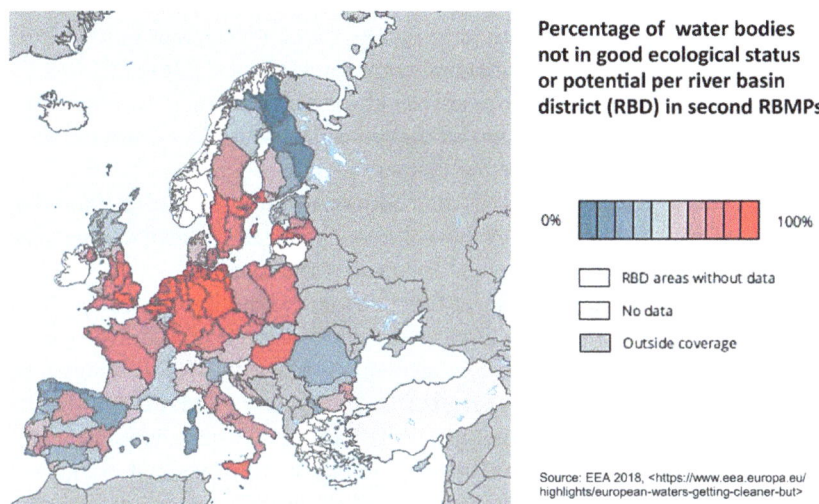

Fig. 3.35: Percentage of water bodies not in good ecological status or potential per river basin district (RBD). Source: (EEA 2018) <https://www.eea.europa.eu/highlights/european-waters-getting-cleaner-but> [accessed: 7.08.2018]

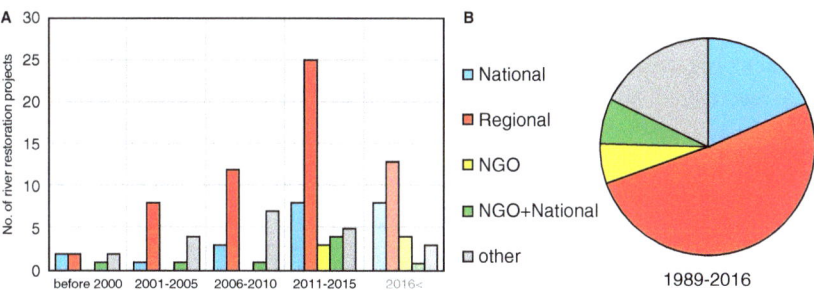

Fig. 3.36: The number and types of entities involved in the projects on regeneration of rivers in Europe from 1989 to 2016 (Szałkiewicz 2018)

---

193 The research by E. Szałkiewicz, S. Jusika i M. Grygoruka (2018) proves that 56 % of the 119 researched European projects of river revitalisation were implemented by

The **example of the Rhine** (cf. Chap. 2.3) provides a good illustration of the difficulties relating to regeneration of big, engineered waterways resulting from the necessity of trans-boundary cooperation. In 1950 in Basel, International Commission for the Protection of the Rhine (ICPR) was founded, including representatives of Switzerland, France, Germany, Luxemburg and Holland. The Commission began work 13 years after signing the Convention in Brno in 1963. The difficulties in reaching multilateral compromises show in the fact that the first agreement which allowed to assess the Rhine contamination, was concluded in 1970. Six years later, Convention for the protection of the Rhine against chemical pollution was signed; while the first sustainable programme of actions for the Rhine was developed in 1987[194], after Sandoz disaster in Basel in 1986 when as a consequence of a fire about 30 tons of toxic chemicals polluted the river. The aim of the programme was to renew and protect the ecosystem of the river. All the decisions influencing ecological functions and simultaneously concerning the river governance including navigation, flood protection and hydropower, required coordinating plans and implementations by the Rhine states (Van Dijk et al 1995). As a result of raising the environmental standards, limiting the sewage discharge, building numerous sewage treatment plants and implementation of monitoring water quality, the condition of the river has improved which enabled further revitalisation actions aimed at biodiversity recreation, including reintroduction of salmon and eel (Kuijpers 1995).The purpose of the programme was also to create ecological system, along the Rhine and its tributaries, consisting of a few big node areas (surface 1,000–6,000 ha) and connecting the wildlife corridors (Van Dijk et al 1995). The enhancement of migration patency and restoration of water-dependant ecosystems as well as connecting the river with flood plains, previously separated from the valley[195], were the objectives of another programme "Rhine 2020". From 2000 to 2012, 122 km² of flood plains were reactivated, 80 reservoirs and watercourses were connected with the Rhine

---

the assigned entities and interested parties but not as part of structural policy of river reconstruction on a bigger scale. It proves that the majority of European countries do not have integrated plans on river revitalisation.

194 Source: Convention on the Protection of the Rhine, 12 April 1999. <http://ec.europa.eu/world/agreements/prepareCreateTreatiesWorkspace/treatiesGeneralData.do?step=0&redirect=true&treatyId=634> [accessed: 22.02.2020].

195 As a result of engineering and embankments, the Rhine lost approx. 85 % of flood plains.

again, while the structural diversity of the banks has increased along 105 km long section of the main Rhine course (ICPR 2013).

The Rhine is the busiest European waterway[196], therefore, it is difficult to find the compromise between the requirements of navigation and the needs of re-naturalisation (More: Cioc 2002; Knepper 2006). Nevertheless, the "Living Rhine" project, approved in 2002, included realisation of 15 model re-naturalisation investments in the area of Germany, consisting in i.a. recreating the riparian forests and meadows, connecting old river beds, creating shifting gravel islands and pro-ecological redevelopment of embankments. The retention capacity of the valley has been raised by relocation of embankments or flood polders (e.g. Altenheim and Sollingen/Greffern – 1,000 ha of the total surface). The Rhine regeneration had fundamental significance for the development of the region and particular cities of North Rhine-Westphalia, especially the recreation of the identity and activation of urban riparian areas of Bonn and Köln.[197] The "Living Rhine" project, realised in France, covered 16,000 ha[198] and was mostly orientated towards the reconstruction of the connections of the river with the former riverbeds of the Rhine, creating spaces of controlled floods and developing methods of retaining crucial riparian habitats.

The quality of the Rhine environment largely depends on its tributaries condition, therefore, the re-naturalisation actions have been undertaken in the whole basin, e.g. by the Neckar River[199], the Enz, or the Emscher rivers. Also, the revitalisation of post-industrial terrains is frequently accompanied by the attempts to regenerate small watercourses. In the Ruhr district, within the framework

---

196 Regular navigation for up to 2,000 ton capacity ships is possible over the length of 868 km and the total length of waterways in the Rhine basin reaches 3,000 km. The shipments between Rheinfelden and the Dutch border (over the length of 622 km) amount to 200 m tons. Also, there is tourist-leisure transport on the Rhine, especially in the section Mainz-Köln. PWN <https://encyklopedia.pwn.pl/haslo/Ren;3967052.html> [accessed: 5.09.2018].
197 The book titled *Riverscapes. Designing Urban Embankments* edited by Christoph Hölzer, Tobias Hundt, Carolin Lüke and Oliver G. Hamm (2008) presents i.a. 16 selected realisations of development and activation of riparian areas in the North Rhine-Westphalia.
198 The re-naturalised areas constitute part of Natura 2000 terrain called Rhine-Ried-Bruch whose surface area is 24,000 ha.
199 By 2011, in Stuttgart and its suburban zone, over 11 km of watercourses have been re-naturalised, restoring the possibilities of meandering and natural character of the watercourses (e.g. the areas by the Neckar and the Feuerbach Rivers). Data from the project: REURIS 2009–2011.

of the Internationale Bauausstellung (International Architecture Exhibition), already from 1989 to 1999 the Emscher river system was reconstructed (within the IBA Emsher Park project) and Duisburg Nord Landscape Park was created (project by Latz + Partner, 1991). In 2005 "Emscher Landscape Park 2010" plan, covering 457 km², was developed, and the recreated and re-naturalised Emscher River became the axis of regional landscape layout, integrating industrial-technological, service and housing investments (Seltmann 2007). Streamlining water governance included: modernisation of sewage systems (400 km of sanitary collectors along the river, 8 sewage treatment plants) and reduction of rain runoff[200], i.a. through financial encouragement to disconnect downspouts from private areas[201] (SWITCH 2006–2011).

Reconstruction of urban watercourses potential

"Development of cities has never been adjusted to the natural watercourses but it was rather the result of their modelling, and subsequent conscious choices eventuated in their present image" (Nyka 2013). The decisions were relevant to the level of knowledge and the requirements prevailing at the time when they were made. The described processes relating to draining wetlands, engineering and canalising watercourses, have created inimitable cultural landscape of cities and foundations for economic development but also deprived cities of ecosystems and retention capability, which caused water pollution and escalation of flooding phenomena.

The return period determined new civilisation priorities and new challenges which have to be addressed in all city structures. Regeneration of urban sections of rivers is much more difficult and transformation possibilities – limited by the existing built environment and infrastructure[202]. On the basis of several

---

200 The document titled *Future Convention for Stormwater in the Emscher Catchment* obliged 17 cities located in the catchment area of the Emscher River to reduce the rain runoff by 15 % over 15 years (by 2020).

201 In accordance with the German law, every owner of a property has to pay "rain charge" of € 0.80 per 1 m² of a sealed area yearly. Moreover, 70 % of the expenses of infrastructure, which is indispensable to perform the disconnection e.g. small retention and infiltration systems, are incurred by Emscher Genossenschaft (Cooperative Society) while the remaining 30 % – by the owner.

202 Already in the late 1970s, waterfront renewal actions were undertaken, which also included environmental regeneration, e.g. river valley renewal in Glasgow or environmental improvement in the river valleys of Greater Manchester (Clouston et al. 1979). The principles and methods of rivers' rehabilitation were described by e.g. de Waal et al. (1998).

Fig. 3.37: The Isar river before the redevelopment (on the left) and after the re-naturalisation (on the right). Photo by Daniela Schaufuß; source: Munich city, <https://climate-adapt.eea.europa.eu/metadata/case-studies/isar-plan-2013-water-management-plan-and-restoration-of-the-isar-river-munich-germany> [accessed: 7.09.2018]

examples, the main directions of natural regeneration in cities were illustrated, ranging from urban watercourses re-naturalisation and creating riparian parks, through examples of uncovering canalised streams to implementation of city-wide strategies of blue-green systems.

The most significant European examples of **re-naturalisation** of urban sections of bigger rivers include the **Isar in Munich** (Fig. 3.37). The river was engineered in the 19th century because of the flood risk and in the 20th century hydroelectric barrages were constructed on its course. The "Isar Plan" was developed in 1995, although the initial preparation had already begun in the 1980s. The Isar re-naturalization project in Munich[203] was carried out from 2000 to 2011. The engineered river was brought to almost natural condition in the section of 8.3 km. The most important objectives of the re-naturalisation included (Bańkowska et al 2010):

- giving the river more natural image and integrating it with the landscape;
- improvement of flood control;

---

[203] The cost of the project realisation in total amounted to 35 m Euros: € 28 m – the construction, 7 m – remediation of the contaminated terrains and removing the weapon remaining since the World War II. The expenses were covered by the Bavarian State Government (55 %) and the city of Munich (45 %). More: Isar-Plan – Water management plan and restoration of the Isar River, Munich (Germany) (2015), <https://climate-adapt.eea.europa.eu/metadata/case-studies/isar-plan-2013-water-management-plan-and-restoration-of-the-isar-river-munich-germany> [accessed: 7.09.2018].

- regaining the river continuity through redevelopment of the existing transverse development and/or constructing fish ladders;
- enhancement of the ecological connection of the river with the floodplains through redevelopment of the banks and the embankments;
- improvement of ecological conditions and habitats diversity – creating and extension of river structures necessary for existence and development of wildlife;
- increase in morphological dynamics of the river, providing it with the suitable space and continuity of sediment transport;
- raising the amount of water remaining in the river bed (change in water governance through hydropower);
- enhancement of recreation conditions and access to the river (concentration of leisure services in suitable places and decreasing leisure pressure in others) i.a. through special shape of the banks;
- improvement of water quality.

The re-naturalisation action covered the riverbed, the bank zone as well as inter-embankment zone. The river bed was widened from 50 m to 90 m and the biggest obstacles posing the flood risk were removed. In some places the river bank enforcements were dismantled in order to provide the river with the possibility of unrestrained shaping of the riverbank line (own dynamics), thanks to which fluvial gravel deposits, backwaters, stagnating tributaries, which are invaluable biotopes for numerous fauna species, occurred. In the places of barrages, bridges or dense development, the throughput was restored by constructing modern fish ladders (Wawręty 2004). The concrete barrages were replaced by flat, stony ramps. Riparian meadows were created and the slopes of scarps were flattened (approx. 1:10), which enhanced their accessibility for the users and the plants (Bańkowska et al 2010). As a result of the transformations, quasi-natural river valley was created, enriching the landscape and providing natural and leisure advantages.

The German examples which are worth mentioning include: parts of the Elbe in Hamburg, the Emscher river in Deinhauser Bach and Castrop-Rauxel near Dortmund, the Leine in Hannover, the Pegnitz in Nuremberg, the projects on the Neckar river in Stuttgart and Ludwigsburg, the Neckar and the Elz in Mosbach, on the Kaczawa and the Weidigtbach in Dresden, or the Wandse in Wandsbek.

Numerous projects on urban rivers and streams regeneration were realised in Great Britain, i.a. on the Quaggy river and the Chinbrook Meadows (LB Lewisham) in London, redevelopment of the Marden river in Calne centre, or the Skerne and Ravensbourne in Darlington[204]. In Czech Republic sections of

---

[204] In Great Britain a few projects on rivers re-naturalisation were realised, 15 of which can be found in *Manual of River Restoration Techniques,* published by the River

the Botič and Krůtecký streams in Prague were re-naturalised, the mill stream in Chrudim, parts of the Loučná River in Litomyšl and the Sázava in the city of Havlíčkův Brod and many others (REURIS 2012).

In Poland the examples of natural regeneration of urban rivers are not plentiful[205]. Gradually, the natural assets of the rivers of Łódź are being discovered and the revitalised valley of the Sokołówka[206] became an axis of attractive public spaces and the system of rain water governance in Łódź. Other rivers, worth mentioning, are Rakówka in Bełchatów and Ślepiotka in Katowice. The example of the tiny river of **Ślepionka in Katowice** (8 km long)[207] excellently features typical problems of small watercourses in Poland: ashamedly hidden at the rear of the buildings, collecting storm runoff and pollution from the highly urbanised catchment area. Before the transformation, the river section was hard to reach, neglected, covered with invasive plants. The main objective of the project was recreating the blue-green corridor of the river valley and creating open leisure-educational space there. The regeneration also allowed to change the model of rainwater governance in the city. Half-natural wetland was created which enhanced the retention capacity of the valley, made it possible to create water and water-land habitats as well as improved the self-purification capabilities of the watercourse. Buffer zones, which protect the river from contamination, were developed with the use of bio-engineering methods. Biological biodiversity increased through the use of indigenous plants, and habitats typical for dry and riparian forests as well as restoration of wetlands. The project managed to reach a compromise between the requirements of the environmental protection and the inhabitants' needs. Not only was the biological recreation of the valley

---

    Restoration Centre (RRC n.d.) <https://www.therrc.co.uk/manual-river-restoration-techniques> [accessed: 22.02.2020].
205 More re-naturalisation initiatives are undertaken in the extra-urban areas e.g. re-naturalisation of the influx section of the Kwacza River (Pomeranian Voivodeship), wetlands called "Pyszka" on the Pysznica river or a big project on re-naturalisation of the hydro-graphic system in the middle basin of the Biebrza valley.
206 The valley of the Sokołówka river in Łódź was regenerated within the frame of international SWITCH 2006–2009 project (*Sustainable Water Management Improves Tomorrows Cities Health*). The actions included i.a. re-naturalisation of the watercourse bed and creating leisure-retention reservoirs (Zgierska reservoir, "staw Teresy" - Teresa's pond, Żabieniec).
207 The re-naturalisation of Ślepotka was performed from 2008 to 2012 within the REURIS project, financed by Central Europe and European Regional Development Fund (ERDF) and the funds of the city of Katowice. Coordinator: GIG in Katowice.

provided but also its accessibility enhanced (steps, paths, catwalks) as well as places of integration and leisure created (amphitheatre, arbours, playgrounds). Educational paths and information boards presenting the role of the plant habitats contributed to the inhabitants' knowledge and greater concern about the environment. The involvement of the society appeared to be one of the most important success factors in the project realisation (REURIS 2012). Workshops and meetings with the residents, experts and decision-makers led to consensus on the project priorities and development of a partnership model in the processes of river valleys revitalisation, satisfying various interests and providing cohesion of the objectives.

Protection and regeneration of urban sections of river valleys as well as management of flood risk are closely related to rainwater governance and the use of "green infrastructure". Creating green buffer zones of riparian greenery and linear parks provided biological protection of watercourses and flood risk management as well as enhancement of greenery structures and leisure conditions in cities.

Uncovering small watercourses became possible after retrofitting of urban sewage systems and improvement in water quality. Rivers and streams buried under ground have been again brought to the surface of American (e.g. the rivers of Los Angeles), Asian (e.g. Seoul, Suwon, Sapporo) and European cities (e.g. the Hague and Breda, Ghent, in Leipzig, Chambéry, Arhus, or in Chech Chrudim and Prague).

Uncovering or reconstructing watercourses allows to regain their cultural identity and biological life but, most importantly, to reduce the heat island effect, manage rainwater and improve living conditions in cities owing to creating new leisure spaces for the inhabitants.

Comparing numerous examples of public spaces before and after the regeneration of small watercourses in European and Asian cities (Fig. 3.38–3.43)[208] clearly shows the changes in visual, functional and environmental quality: from car communication-oriented to user- and nature-friendly ones.

---

208 The presented comparisons come from URB-I: URBAN IDEAS portal; BEFORE /AFTER CATEGORIES; RECOVERY OF STREAMS <http://www.urb-i.com/recovery-of-streams> [accessed: 11.08.2017].

Environment regeneration

Fig. 3.38: Anam-ro/Seognbukcheon, Seoul, South Korea (URB-I n.d.). Above: the view before uncovering the watercourse. Below: after the restoration

Fig. 3.39: Suwon cheon, Suwon, South Korea (URB-I n.d.). Above: the view before uncovering the watercourse. Below: after the restoration

Fig. 3.40: Shinan-gil, Daejon, South Korea (URB-I n.d.). Above: before, below: after the changes

Fig. 3.41: Noordwal/Veenkade, the Hague, Holland (URB-I n.d.). Above: before, below: after the changes

Environment regeneration 127

Fig. 3.42: Sint-Jacobsnieuwstraat, Ghent, Belgium (URB-I n.d.). Above: before, below: after the changes

Fig. 3.43: Quai du Jeu de Paume, Chambéry, France (URB-I n.d.).Above: before, below: after the changes

A spectacular example of natural and cultural restoration of a watercourse can be found in the project on the reclamation of the **Cheonggyecheon stream in Seoul** (project by SeoAhn Total Landscape), which was previously canalised and covered with a multilane motorway[209] in the 1970s. Over time, it began to disappear even from the inhabitants' memory. Drastic deterioration of living conditions in the city, caused by intensification of traffic and greenery shortage, eventually prompted the authorities to undertake comprehensive revitalisation actions. In 2003 Seoul mayor, Lee Myung-Bak, made a controversial decision to liquidate almost 6 km of the motorway in favour of giving the river back to the city along with all the benefits of its climatic, natural, landscape and leisure advantages. From 2003 to 2005, in the centre of the capital, in the place of the former "river of cars", over 8 km blue-green strip of public space was constructed, lining the uncovered Cheonggyecheon stream (Fig. 3.44), which soon became a showcase of Seoul and the entire South Korea. The realisation of the investment entailed serious problems, though. In addition to traffic holdups and high construction costs[210], the social expenses were estimated at $ 1,900 m. Traders' mass protests made the authorities implement compensatory mechanisms allowing for partial loss compensation. The process caused gentrification of the district and increase in the rent and value of the neighbouring properties (30–50 %). In many respects, the *Cheonggyecheon River Restoration Project* was a model project of a district revitalisation[211] based on a stream restoration. Its realisation lasted only 2 years (2003–2005) and brought impressive balance of social, economic, space and environmental benefits. The positive changes worth mentioning include the effect of reducing heavy traffic in favour of public and pedestrian communication, which resulted in the fall of noise and exhaust fumes pollution by 35 %. In comparison to year 2003, the restoration of the river valley allowed to reduce the urban heat island effect and air the city as well as boost biodiversity by 639 % (Cheonggyecheon… n.d.). It also had considerable significance for regaining cultural identity of the part of Seoul – the river appeared again in

---

209 More: *Cheonggyecheon River Restoration Project: The Restoration of Environmental, Social & Economic in Seoul* <http://www.nclurbandesign.org/tag/the-cheonggyecheon-stream/> [accessed: 17.04.2015].
210 The expenses were estimated at $367. Eventually, they were almost three times higher.
211 The investment was coordinated and supervised by the *Cheonggyecheon Restoration Project Headquarters*. A team of negotiators was also appointed to solve social conflicts – *Citizen's Committee for Cheonggyecheon Restoration Project* as well as *Cheonggyecheon Restoration Research Corps* – a team of experts responsible for developing the details of operational plans and technical control (Zujewski 2014).

public consciousness along with the former bridges, places of leisure, play, rituals and art. The public river space is used by approx. 90,000 people daily, and the city centre has gained a new, inhabitant- and nature-friendly image (Zujewski 2014).

Source: *Seoul Metropolitan Government:*   BEFORE   AFTER

Fig. 3.44: Seul, uncovering the Cheonggyecheon stream. On the left: the view before, on the right: after the stream restoration (LAF n.d., Cheonggyecheon Stream Restoration Project)

Houtan Park (Fig. 3.45), constructed for Expo'2010 in **Shanghai** (China) by **Huangpu** river (project by Turenscape)[212], is an exquisite example of a buffer park, serving waterfront regeneration, water cleaning, flood protection and recreation.

The narrow strip of a degraded area along the Huangpu river (surface area: 14 ha, length: 1.7 km, width: 5–30 m) was formerly owned by a steel factory and a shipyard. The layers of agricultural and industrial past as well as the vision of future eco-civilisation overlap in the re-naturalised landscape. Linear park consists of artificial wetlands, serving as a living machine purifying the contaminated water of the Huangpu river. Cascades and terraces enable oxygenation, removal of bio-genes and reduction of sediments whereas the paths network encourages the users to benefit from recreational and educational facilities. The park also performs a function of anti-flood buffer. The wetlands constitute the river expansion space in the zone between 20- and

---

212 The project received ASLA Professional Awards in 2010. The description based on: Award of Excellence. Shanghai Houtan Park: Landscape as a Living System, Shanghai, China. Project: Turenscape, China and Peking University Graduate School of Landscape Architecture, Client: The 2010 Shanghai Expo Bureau, China, source: <https://www.asla.org/2010awards/006.html> [accessed: 12.11.2018].

1000-year water extent. Artificially created wetland fluvial terraces, ecological system of flood control, indigenous species selection and urban crops are part of a more comprehensive strategy of the valley regeneration aiming to purify the river water.

Fig. 3.45: Shanghai, Houtan Park by the Huangpu river. On the left: the view before, on the right: after the waterfront regeneration (LAF n.d., Houtan Park)

The decision to uncover and regenerate a watercourse was already taken in the 1980s in **Leipzig**. Owing to social initiative, seven 4 km long sections of the **Pleißemühlgraben Canal** (Fig. 3.46) in the southern, western and north-western parts of the city centre of Leipzig were reconstructed from 1996 to 2007. The works included: uncovering a mill stream, developing the neighbouring green areas, creating a water playground as well as a catwalk for pedestrians, along with floating gardens on the Nonnenmühlwehr weir. The objectives of the project covered the following aspects (REURIS 2012; Lange and Nissen 2012):

- ecological: determined by the Water Framework Directive of the EU;
- economic: raising the value of the waterside properties and tourist attractiveness as an element of the strategy of waterways activation in Leipzig region as well as modernisation of the Leipzig water node;
- social: enhancement of recreation and leisure conditions for the inhabitants and ecological education (engagement of schools and non-governmental organisations as well as the residents, during and after the project implementation);
- spatial: cohesion of the historically meaningful elements (e.g. the development, the canal embankment forms: vertical, steps) and modern solutions of small architecture (railings and lighting).

Environment regeneration 131

Fig. 3.46: Leipzig, regeneration of the Pleißemühlgraben canal. On the left: the views before revitalisation. On the right: uncovered and regenerated sections (Photo by A. Januchta-Szostak, photos from the 1990s before revitalisation: REURIS 2012)

The undertaking was realised in accordance with the regulations of public-private partnership[213]. Education and engagement of the society appeared to be crucial success factors. In the primary, secondary schools and universities information and didactic programmes on ecological education and environmental communication were implemented. As a result, the investment not only improved the image and living conditions in Leipzig but also raised the inhabitants' awareness of the environment quality issues.

Regeneration of the **Manzanares river in Madrid** (Fig. 3.47) is an example of a different kind of action which consisted in raising weir gates and enabling the river to shape natural riverbed forms within embanked, entirely artificial banks (photo and comment by P. Nawrocki, WWF Poland). Although the river accessibility for the inhabitants is still low, its biocenotic and landscape values have considerably increased.

---

213 Revitalisation cost of 1,100 m section, opened in 2007, amounted to € 9.5 m. The works were financed by the city of Leipzig (1/3), subsidies from the Free State of Saxony and two German foundations for environment – *Deutsche Bundesstiftung Umwelt* as well as *Allianz Umweltstiftung* (1/3), and the third part by private owners of riparian areas.

Fig. 3.47: Madrid, the Manzanares river – natural regeneration of the riverbed (photo by P. Nawrocki)

Blue-green networks

The tradition of pro-ecological development of river valleys dates back to the end of the 19th century when Frederick Law Olmsted designed a system of river parks called Emerald Necklace in Boston (1894). A hydrographical system also became the axis of Poznań greenery wedges, created with the use of the river and its tributaries: the Bogdanka[214], the Cybina, the Główna and the Junikowski Stream (project by W. Czarnecki, A. Wodziczko, 1930–1934). However, in majority of big European cities, the rivers require restoration and reconstruction of ecosystems.

---

214 The Bogdanka river is the axis of the western greenery wedge in Poznań. It is diversified by numerous lakes (Kierskie, Strzeszyńskie, Rusałka) and ponds in beautiful 19th century Sołacki Park. The valley serves important social, landscape and ecological functions. Regrettably the section from Wodziczko Park to the Bogdanka confluence with the Warta has been canalised which has broken the natural and leisure connections between the Warta valley and its urban tributary (More: Januchta-Szostak 2011a).

Environment regeneration

Even fragmentary revitalisation of urban sections of river valleys has local significance but it does not facilitate restoration of the system of ecological corridors and does not generate a synergy effect which occurs in the case of integrated hydro-graphic systems. The most important attribute of the "blue-green systems" is the continuity of the linear layout of greenery and water[215]. This aspect is not only crucial for the enhancement of their public accessibility and leisure use but also for increasing biodiversity of the habitats and improvement of the rivers' self-purification capability.

Internal greenery system of **London** comprises scattered spots which are not linked into corridors along river valleys. Rivers in the city served mostly transport and sewerage functions therefore they were deprived of their ecosystems. The scale of degradation of greenery structures and water prompted the city authorities to develop a plan of reinforcing and reconstructing eco-corridors (Fig. 3.48) as well as revitalising urban rivers. In London, since 2004 a wide-reaching programme on urban rivers revitalisation called **The Blue Ribbon Network** (BRN, cf. Chap. 4.3) has been implemented. BRN is integrated with spatial planning as well as water and city environment strategies: *London Water Strategy* (Mayor of London 2011), *London Environment Strategy* (Mayor of London 2018). Its main goals, formulated in *London Rivers Action Plan* (River Restoration Centre 2009) include:

1) improvement of the flood risk management with the use of more natural processes;

2) mitigating negative climate change consequences;

3) accessibility of the natural environment to the inhabitants through urban-natural revitalisation;

4) enhancement of leisure conditions and life quality;

5) reinforcement of wildlife habitats.

Integrated actions undertaken in Poland can be illustrated by the example of the **Blue-Green System in Łódź**[216] (Fig. 3.49) whose objective is to connect

---

215 Linear parks are increasingly more popular in Spain (parks in Madrid, Barcelona, Valencia, Saragossa, Malaga, Cordoba or Seville).

216 The concept of Blue-Green System was developed in European Regional Centre for Eco-hydrology (ERCE) under the aegis of UNESCO in Łódź (the authors of the concept: M. Zalewski, I. Wagner, 2009), within the project IP 6 PR UE, GOCE 018530 2006-2011. Source: The report on the meeting "Blue-Green System" – developing

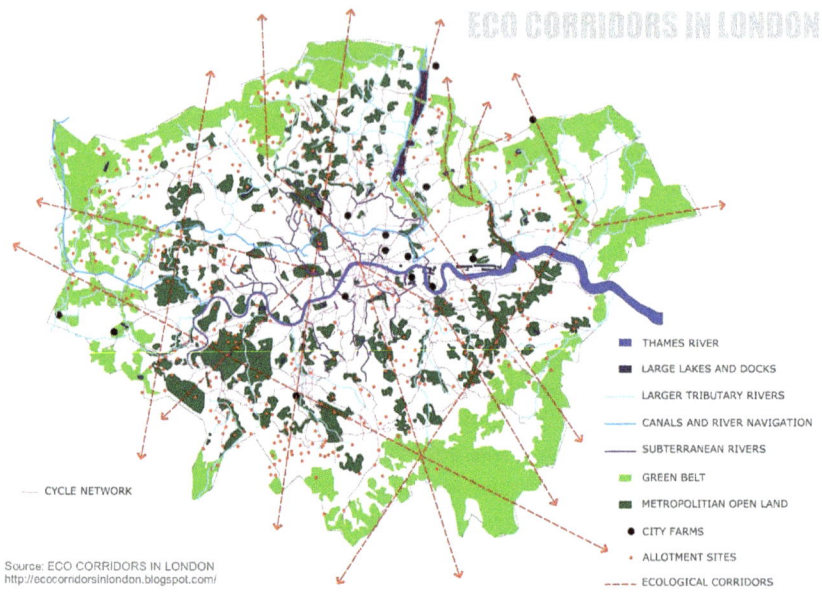

Fig. 3.48: Blue-green structure of London requires restoration of valley eco-corridors. Source: ECO CORRIDORS IN LONDON, <http://ecocorridorsinlondon.blogspot.com/> [accessed: 12.11.2018]

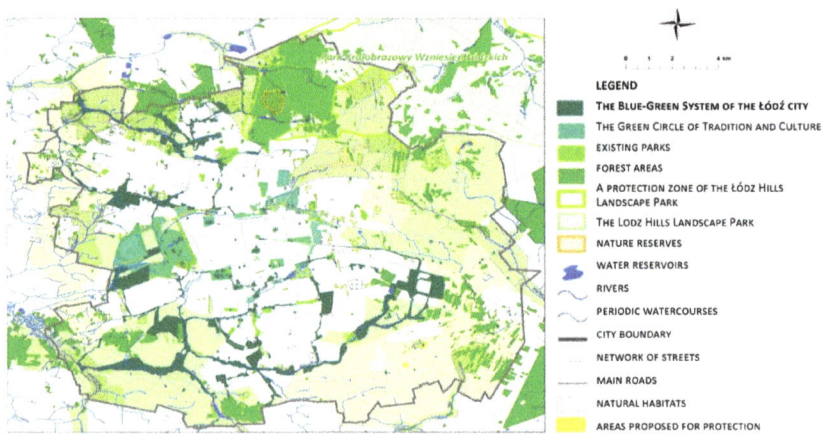

Fig. 3.49: Location of the Sokołówka valley with components of Blue-Green Network in Łódź in the background (in accordance with Zalewski 2010). The map on the basis of: Krauze et al 2010, p. 35

the hydro-graphical system with the natural system of the city and the region. The process of scientists and decision makers' integration of actions on the issues concerning green and blue infrastructure intensified in 2006 when stakeholders panel Learning Alliance (LA) was appointed. The LA team formulated two amendments to the *Study on Conditions and Directions of the City of Łódź Spatial* Development (developed in 2010) concerning sustainable rainwater governance and implementation of the Blue-Green System (BGS) as an essential element of city spatial development. In 2012, on the basis of the act of the City Council, the BGS became part of city strategy and the first stage of its realisation was re-naturalisation of the Sokołówka river. The Blue-Green System is a project giving foundations for functional, economic as well as logical and friendly layout of urban space, which benefits the inhabitants in a number of ways. They include i.a.: increased retention and purification of rainwater in the landscape simultaneously preventing droughts and floods; enhancement of microclimate and air quality; raising resilience and lowering the expenses of urban greenery areas maintenance; adaptation to climatic changes; facilitating leisure and ecological public transport; enhancing the attractiveness of the urban spaces for the inhabitants and investors (Wagner et al 2013).

Investments concerning the construction of the blue-green infrastructure, including natural regeneration and rivers re-naturalisation, last long and are not nearly as spectacular as urban waterfronts transformations. They do not generate short-term profits[217], tend to be surface- and assets-consuming, entail many risks and uncertainty factors but constitute a necessary stage of restoring natural processes of water circulation and purification in catchment areas. Thus omitting the stage and concentrating mainly on waterside areas aesthetics reminds using make up instead of treatment.

3.4 Water governance

Over half a century has passed since the Committee of Ministers adopted the *European Water Charter* (1968), introducing fundamental regulations on

---

   urban space for the enhancement of life quality and sustainable development of Łódź, 28 May 2009; http://switchlodz.wordpress.com/blekitno-zielona-siecblue-green-network/ [16.05.2011]
217 Whereas in a longer-term perspective "every £1 invested in river renewal brings £7 long-term new benefit" (cited in Oates 2012).

quantitative and qualitative water protection. At that time it was such a revolutionary declaration that the basic rights of water (like the right to purity and self-purification through the connection with the catchment vegetation, the right to circulation and recovery of resources) began to be respected only after 30–40 years[218]. Its main postulates were reflected in the **Water Framework Directive** documents (2000/60/EC). The right to overflow and flood beyond the administratively determined banks was "returned" to the rivers only under the **Floods Directive** (2007/60/EC), which acknowledged the flood as an unavoidable natural phenomenon. Flood risk obviously poses socio-economic problems and its reduction is one of the three main objectives of water governance along with water resources protection and satisfying justifiable social and economic water needs (Nachlik 2006). Nevertheless, the essential change in the approach consists in skilful flood risk management instead of "fight against the flood" which was the prevailing attitude in the conquest period.

The process of the *Integrated Water Resources Management* (IWRM 2000), consistent with WFD, is the foundation for shaping the water policy of the EU states, enabling profitable and sustainable governance without the necessity of disturbing ecological balance of the ecosystems. Theoretically, the structure of the process[219] is based on the principles of social justice (providing all the users with access to water whose quality and quantity are determined), economic efficiency and ecological balance (IWRM 2000). Practically, European river ecosystems have not regained balance. Considerable areas of basins are still being urbanised and water governance in cities follows its own principles. Meeting the rising water demand in urban agglomerations is the priority issue while the needs of ecosystems are less urgent. No wonder, the processes of urban rivers regeneration face numerous organisational, financial and ownership barriers.

---

218 Only transposition of directives to law making institutions of the EU member states allowed to sanction them locally.
219 The *OECD Principles on Water Governance,* 2015, emphasise productivity and efficiency of water governance as well as trust and involvement of various stakeholders in shaping water policy.

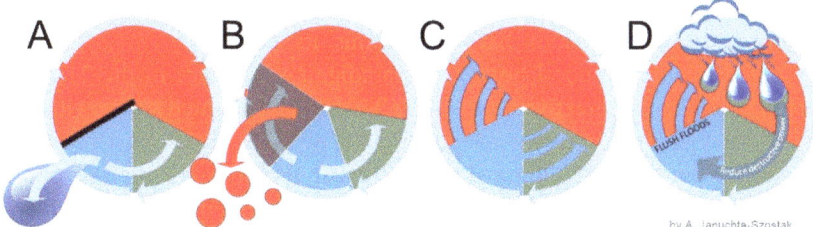

Fig. 3.50: Flood protection strategies: A. Keeping water away from people (e.g. embankments), B. Moving people away from water (e.g. ban on floodplains development), C. Coexistence with water (e.g. non-defensive urbanisation methods), D. Preventive actions in the entire catchment area. Developed by A. Januchta-Szostak

Flood risk management strategies

A representative variable in relation humans-river is the approach to flood risk. In the respect period, safety was provided by location of the settlements in the areas above floodplains, far from the reach of high overflows. During the conquest period, though, urbanisation pressure and benefits resulting from close river vicinity imposed accepting a more risky strategy of **"keeping water away from people"** (Fig. 3.50A), especially that flood precautions (embankments, dams, barriers, reservoirs, polders, relief canals etc.) made it possible to fight the power of the element and control the water flow. Narrowing floodplains in one place, however, increased hazards downstream due to acceleration of flow and rising flood wave. Despite this fact, **defensive strategies**, focused on limiting the extent of flood hazard areas, were prevailing in Europe and the USA[220] in the 20th century and still dominate in Poland. Abandoning this strategy is not easy and sometimes unattainable because of considerable development of the floodplains. Entering water territory has been a common policy, especially in cities suffering from deficit of investment areas. The flood plains that were once protected by technical systems, have been intensively developed, which has resulted in increasingly heavier losses in the event of damage to the embankments or other hydrotechnical structures. Another issue is the high financial and environmental cost incurred in maintaining flood protection infrastructure. However, taking into consideration the scale of the catchment areas transformations and

---

220 The United States Flood Control Act of 1936 set the direction for structural defences as the primary flood protection strategy in many other countries.

European rivers engineering, many of which have lost over 80 % of the original floodplains, it seems obvious that water "claims" its space.

The change in the flood protection paradigms took place in the 21st century due to Floods Directive (2007/60/EC), which highlighted the inevitability of flood risk and the necessity to reconcile the need for security with development demands. Therefore, the goals have been shifted from flood protection to limiting its negative effects of floods and risk management. According to FD "flood risk means the combination of the probability of a flood event and of the potential adverse consequences for human health, the environment, cultural heritage and economic activity associated with a flood event" (2007/60/EC, art. 2, p.2.). The scale of these negative effects depends on: the **extent of the flood** (**flood hazard**), the state of use of the affected area (**exposure**), the **vulnerability** of the development elements to the threat, and the ability of local communities to counteract the threat and liquidate the effects of the disaster (sensitivity). Contemporary flood mitigation strategies are based on the modification of these three variables (KZGW 2013):

- **flood hazard** (determined by the probable extent of floods) that can be limited on the one hand by technical means such as levees, embankments, reservoirs, relief channels or riverbed regulations (Fig. 3.50A). On the other hand the sources of flood threats can be reduced by developing natural and artificial retention and rainwater management (Fig. 3.50D);
- **exposure** (understood as buildings, objects and communities located in hazardous areas), which can be minimized by reducing investment in floodplains, mainly through bans or restrictions on building development, or setting special conditions for the construction of objects and by purchasing land and providing compensation. The **"retreat and respect" strategy** is associated with **"moving people away from the water"** (Fig. 3.50B), but also adapting possible development to eco-hydrological conditions;
- **vulnerability** (defined by the preparation of objects and people for floods), which can be reduced by using structural methods (e.g. protection of building and land development against floods) and non-structural, such as: flood insurance, early warning systems and response to floods, awareness of residents and education about prevention and dealing with floods. The **"life with a flood" strategy** (Fig. 3.50C), focused on reducing vulnerability to floods, is a key element of the broader **"coexistence with water" strategy**.

In practice, "we manage neither to keep destructive waters away from people at all times nor keep the people away from destructive waters" (Kundzewicz et al. 2018, p.12322), so the most effective strategy seems to be "**coexistence** with water", which combines actions aimed to limit the exposure ("retreat and respect" strategy) and vulnerability ("living with a flood" strategy). However, it requires not only adaptation of the site development, buildings and their users to periodic flooding and reliable warning and evacuation systems, but also taking into account the impact of any development on the dynamics of water flow and river ecosystems as well as reduce sources of flood hazard. As a consequence of climatic changes weather extremes intensify, threatening with increasingly more fierce sea, river and rainfall floods. Preparation for such changes and mitigating them requires a strategy which exceeds river valley and provides **water retention in the entire catchment area** (Fig. 3.50D).

Comparative studies of flood risk management carried out as part of the STAR-FLOOD project[221] prove that it is necessary to create appropriate combinations of structural and non-structured mechanisms in different locations. Admittedly, floods constitute a hazard only when humans encroach on flood prone areas, but in the reality of urban environments it is difficult to constrain investment pressure on floodplains. In order to increase the cities' resilience, it is crucial to strengthen the three pillars of the system: **the ability to resist** (e.g. through defence mechanisms), **flexibility in absorbing floods and regaining efficiency** (e.g. through spatial planning, disaster management and insurance), as well as **seizing opportunities** in the process of adaptation and transformation (Kundzewicz et al. 2018).

Space for water in regions

The Floods Directive (2007/60/EC) determined a new direction for the development of valleys and riparian areas, considering increase in their retention capacity. Nevertheless, even before its implementation, European countries began to introduce strategic programmes orientated towards expanding space for water, like *Room for the Rivers* (2006) in Holland, or *Making Space for Water* (2005) in Great Britain.

---

221 The STAR-FLOOD project (2012-2016), carried out in six EU countries (Belgium, England in the United Kingdom, France, The Netherlands, Poland, and Sweden), focused on the analysis, assessment and planning policies aimed at reducing flood risks in urban agglomerations throughout Europe. The results are important for policy and law at European, national and regional level and for the development of public-private partnerships. Source: <www.starflood.eu/> [accessed: 18.02.2019].

The British solutions, developed within the framework of the programme called **Making Space for Water** (DEFRA 2005), allowed for flexible, non-defensive systems of flood risk management. The programme focused on i.a. the integration of urban planning with sustainable water governance, both in the areas of river valleys (protection against external threats) and in the catchment areas (increasing rainwater retention in order to reduce urban floods and protect lower located areas). The objectives of the MSW also emphasised integration of two aspects: national priorities to protect the environment and reduce flood risk with the key issues of the local socio-economic policy, especially concerning housing. The initiative called LIFE – *Long-term Initiatives for Flood-risk Environments* includes three programmes on sustainable water and space governance (Baker and Coutts 2009): 1. *Living with Water* – the approach allowing for the development of housing in the flood risk areas, on the condition that the limitations resulting from the increase in flood phenomena occurrence are accepted; 2. *Making Space for Water* – reduction of defensive flood protection systems in favour of creating space where periodical flooding is accepted and controlled not only with the use of technical means but also natural processes; 3. *Zero Carbon* – using all the alternative energy sources in the particular area with the aim to reduce non-renewable raw materials and the use of national transmission networks. Elements of integrated planning were tested in three pilot projects: in Hackbridge by the Wandle river (an upper catchment urban site), Peterborough by the Nene river (a middle catchment sub-urban site), and Littlehampton by the Arun river (a lower catchment rural site). In order to minimise damages, the spatial development was divided into subsequent stages in accordance with the flood risk maps. The suggested solutions made it possible to increase the building intensity and create appealing public spaces as well as on-water leisure areas, simultaneously improving retention capacity and unrestrained access of flood waters to the area. The researched housing estates were provided with sustainable drainage systems (SuDS), which were later recommended as a standard in the British legal act (*The Flood and Water Management Act* 2010).

The Dutch programme called *Room for the Rivers* (2006) addressed the issue of flood risk posed by big, engineered rivers (especially the Rhine), flowing from catchment areas beyond Holland borders. It is a significant fact that the Dutch, suffering from space shortage, acknowledged the necessity of "returning" areas gained for centuries to water and opening them for controlled flooding. The programme aimed at developing methods of raising retention capacity of river valleys and liquidating bottlenecks e.g. through removing polders and barriers, relocating embankments, constructing lateral canals and lowering of floodplains, groynes and bottoms of the rivers. The methods were implemented and tested within the frameworks of 39 projects located at the Rhine estuary by the Wall, the Nederrijn/Lek, the IJssel and partially the Maas rivers. One of them was the redevelopment of a "bottleneck" of the Waal river in Nijmegen (2006) which required relocation of embankments by 350 m from the river and widening the valley by a lateral canal which has a form of meandering semi-natural tributary. Hydro-technical transformations served shaping leisure space for the inhabitants (promenades, cycle routes, beaches) and raising quality of the landscape, which is the second most important goal of the programme, after safety priority. The Dutch model of water governance has been still evolving from the systems which are strictly controlled by means of hydro-technical and sanitary infrastructure to flexible systems allowing for fluctuation, meandering, improving retention capacity and biodiversity of water ecosystems, even of the ones which are closely integrated with the built environment (Januchta-Szostak 2010). The strategy based on preventive actions in the entire catchment area (Fig. 3.50D) was included in National Policy of Spatial Planning of Holland for 2000–2020 *Making Space Sharing Space* (2001) and has been consistently implemented by creating space for water in the basins in urbanised and agricultural or forest areas.

Flood risk management in cities requires considering external threats (e.g. river floods generated in the upper parts of basins) as well as internal risks (e.g. urban floods caused by intensive precipitation and ground sealing) (Fig. 3.51, 3.52). Flood risk maps enable the adjustment of functions and planned development layout to the level of the risk and reduction in flood hazard increase.

The fundamental objectives of flood risk management in cities include (IMGW 2012):

- avoiding increase in flood risk in the future through:
  - maintaining (increasing) present retention capacity of catchment areas,
  - ceasing development in threatened areas e.g. abandoning urbanisation of high flood risk areas > 1 %,
- limiting the present risk by reducing:
  - current flood risk,
  - present development, e.g. adaptation of the development to possible flood occurrences,
  - vulnerability of built environment and communities (moving vulnerability functions beyond the threat zone),

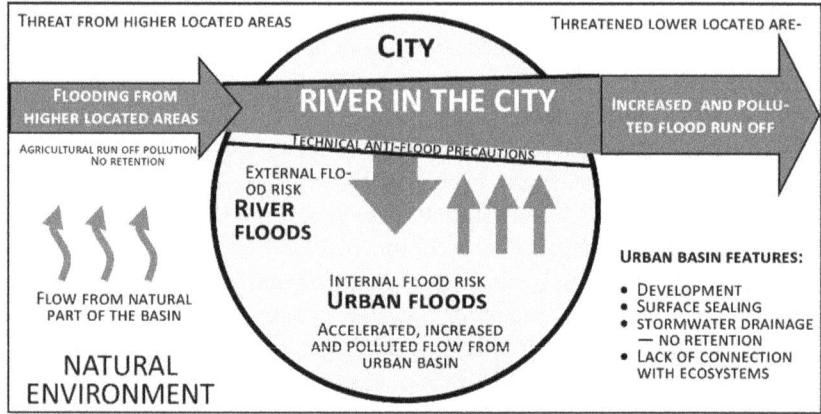

Fig. 3.51: Quantitative and qualitative flood threats in the current catchment area system (developed by A. Januchta-Szostak, partially based on Nachlik 2006)

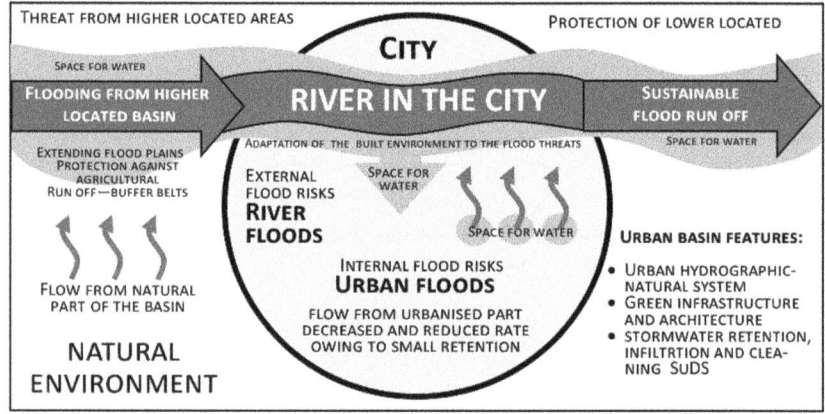

Fig. 3.52: Integrated approach to quantitative and qualitative flood threats in catchment area system (developed by A. Januchta-Szostak)

- mitigation of negative consequences during and after flood owing to enhancement of:
    - forecasting and warning systems,
    - reaction effectiveness of institutions and societies,
    - efficiency of analyses and reconstruction after flood (drawing conclusions, retrofitting).

The actions must be undertaken in basin areas, water regions and individual catchment areas (Romanowicz et al. 2014). In the case of densely urbanised catchment areas, the key actions consist in enabling the threats decrease by reducing the rate and the amount of precipitation runoff from urban areas. Therefore it is crucial that the guidelines on urban catchment retention should be included in the study of spatial development of a commune as well as in the local spatial management plans (Januchta-Szostak 2014b).

Planning under conditions of uncertainty

The methods of natural environment regeneration, presented in the previous subsection, belong to the most efficient tools of surface waters purity enhancement and retention increase but their implementation entails difficulties resulting from the necessity of changing the prevailing technical methods as well as **the concern about losing control**. It was very accurately expressed by Wybe Kuitert: "Water just makes us lose control [...]. Fickle friend or friendly foe" (Kuitert 2008, p. 9). Reconstruction of friendly relations between a city and its river requires then planning under conditions of uncertainty, managing risk instead of making illusory attempts to completely protect against flooding and preparation for hydrometeorological extremes caused by climate change and deepened by the development of urban catchments. If we additionally consider uncertainty connected with the effects of re-naturalisation processes[222], water governance in a city appears to be a **wicked problem**[223], which cannot be solved by one decision, as it is composed of hydrological (B), environmental (G) and cultural (R) aspects including social,

---

222 The realised investments prove that "despite profound pre-project research and analyses, forecasting the effects of re-naturalisation is usually inaccurate. For instance, it is not possible to precisely determine the throughput of the river beds after re-naturalisation, as well as morphological changes in the bed and the valley, or predict the environmental reaction to the changes caused by re-naturalisation. Therefore a crucial element of re-naturalisation process is monitoring the changes in environmental conditions" (Bańkowska et al. 2010, p. 195).
223 "Wicked problems" were defined by Horst Rittel and Melvin Webber in 1973. One right solution to a wicked problem does not exist but there are only better or worse ones; the number of solutions is infinite; no solution can be tested with 100 % certainty; there is no right definition of the problem; defining a problem determines its solution; the problem is not really understandable until the solution is generated; every solution causes problems, each problem can be treated as a symptom of another problem. More: *Wicked problems*, Maciej Lipiec 09/07/2012, http://uxdesign.pl/wicked-problems/> [accessed: 9.09.2018].

functional and economic issues. Planning must allow for numerous variables while the solutions in one field generate problems in another. Nevertheless, flood protection can be combined with maintaining or even restoring natural advantages of rivers and their valleys by using an integrated approach (instead of a sector approach) and flood prevention as well as adjusting the direct protection precautions to the kind of threat (external, internal) and the catchment area profile (Nachlik 2006)[224] (Fig. 3.51, 3.52).

Water quality

Pure water is a necessary condition both for lives of cities and their rivers vitality whereas its quality depends on the state of the environment of the catchment area. Healthy natural environment not only influences the success in revitalisation actions in river valleys but also the possibilities of cities development. Water purity governance considerably improved already at the end of the 20th century, owing to municipal and industrial sewage treatment as well as the control of linear pollution from roads and motorways. The requirements of environmental directives as well as the **Water Framework Directive** (2000/60/EC) have considerably raised the standards[225] of shaping rivers and their valleys as well as concern about the quality of the runoff from the catchment areas, within the scope of the ecological state of watercourses and surface waters. The problems of area pollution from farming areas and rainfall runoff from urbanised areas still remain unsolved along with low self-purification capability of watercourses due to their engineering and canalising in cities. Regrettably, the degraded

---

224 Success in re-naturalisation processes and ecological state of watercourses depend on i.a. the degree of catchment area sealing and the runoff structure resulting from it. In urban areas where 10–25 % of surface is sealed, it is necessary to reinforce the catchment retention (Nachlik 2006) which can be done effectively by using green infrastructure.

225 The quantitative, chemical and biological state of water is the standard of assessment of its ecological condition. In the case of biological condition it is the morphological state of the river bed and its valley. The majority of technical flood protection precautions until now, have intensively interfered into the morphology of the river bed and its valley through hydro-technical development used to engineer and stabilise the river and stream beds. The current solutions aim to search for compromising methods, which guarantee achieving goals of flood protection and acceptable level of ecological condition of water courses (Nachlik 2006).

ecosystems of urban rivers are qualified as Heavily Modified and Artificial Water Bodies (HMWB)[226], which requires only assessment of their ecological potential.

Water pollution in urban rivers was a barrier to spatial development of waterfronts for a long time, especially to raising their leisure values. Only comprehensive actions, connected with considerable reduction of sewage discharge, valleys and riverbeds restoration as well as reconstruction of ecosystems within the tributaries systems, made it possible to enhance the self-purification capabilities of river ecosystems. The subsequent step is preliminary purification of precipitation with the use of riverside buffer parks (RBPs – Januchta-Szostak 2013) and Sustainable (urban) Drainage Systems (SuDS)[227] in the place of precipitation occurrence.

3.5 Integrated urban water management

A city as an ecosystem

The nature was marginalised in urbanised environment. Admittedly, the value of greenery in cities was appreciated already in the second part of the 19th century when the term "ecology" was invented[228] but perceiving the city as an ecosystem began to sprout only in the second part of the 20th century. Early European research within the scope of urban ecology focused mainly on the analysis of greenery structures and vitality of urban biotopes (Sukopp et al 1979, Wittig and Sukopp 1993) whereas in North America a sociological and landscape approach prevailed in the research on urban ecosystems (Pickett et al 1997, Forman and

---

226 The division of homogenous surface water bodies into natural (NWB), heavily modified (HMWB) and artificial (AWB) results from the enties of WFD (2000/60/EC) and water quality classification. Natural water bodies are assessed for their ecological quality while the heavily modified and artificial ones – their ecological potential only.

227 SuDS enable reduction of risk of surface waters flooding, purification of surface runoff, prevention of contamination of rivers and streams tributaries as well as economising on drinking water owing to rainfall water use. Sustainable drainage can be "green" or "grey". "Green" systems use natural plants to purify and store water, while the "grey" systems are based on water-sewage infrastructure like underground reservoirs, retention-infiltration boxes etc. to retain and slowly dispose of water. Both types are effective though the green systems offer extra benefits, increasing ecological protection of watercourses, raising landscape appeal and environment cleanness (*London Environment Strategy*, Mayor od London 2018).

228 The term (Lat. oecologia) was invented by Ernst Haeckel, a German biologist and evolutionist, in 1866 to describe the research on animals and their relations with the surrounding inorganic and organic world.

Godron 1986). Erik Swyngedouw and Maria Kaika (2000) analysed the changes in views on the "nature-city" relation, which have been evolving from decisive antagonising, through attempts to find optimal proportions in model urban planning concepts (e.g. city-garden by Ebenezer Howard, 1898, Arcosanti Paolo Soleri, 1969) to the vision of a city as an ecosystem (Szulczewska 2002).

The ground-breaking publication by Ian McHarg, titled *Design with Nature* (1969), and ten years later – *Nature in Cities* by Ian C. Laurie (1979) laid a sound foundation for numerous visions of cities eco-development: eco-polis (Tjallingii 1995), the currents of landscape urbanism (*Landscape Urbanism,* 1997) by Charles Waldheim (2006), eco-urbanism (*Eco-Urbanism* – Miguel Ruano, 1998), green urbanism (*Green Urbanizm* – Timothy Beatley, 2000), or Emo-urbanism by Charles Anderson (2010). All the "eco-adjectives" indicate **cultural paradigms shift in urbanism** at the turn of the 20th century whose keystones became urban environmental quality and sustainability of natural urban systems which is emphasised by the New Charter of Athens (1998, 2003) and the Leipzig Charter (2007). The breakthrough also occurred in the relations between the city and its river. Apparently, **it is not only the cities that require protection and coexistence with water but also rivers need to be protected from the negative influence of cities.** The return period also means the necessity to consider the integrity of **urbanised (R), natural (G) and water (B) structures as a condition for cities sustainability.**

For the last few years, also in Europe, the research and implementations have been intensified in order to enhance water and environment quality of urban watercourses, natural flood protection precautions as well as integrated water governance in cities. Worth mentioning research projects on the issues include: LIFE (*Long-term Initiatives for Flood-risk Environments*); SMURF (*Sustainable Management of Urban Rivers & Floodplains*), focused on developing rivers and morphological methods of watercourse assessment; URBEM[229] (*Urban River Basin Enhancement Methods* 2004); intercontinental project UNESCO IHP concerning eco-hydrological problems of urban rivers; SWITCH (*Sustainable Water Management Improves Tomorrows Cities Health*); B-SURE (*Building on Small Scale Regeneration of Urban Heritage along Rivers and Canals*), REURIS[230],

---

229 URBEM – a project financed by European Commission within Key Action 4 *City of tomorrow and cultural heritage*, 2004.
230 REURIS project, realised from 2008 to 2012 within the framework of Central Europe programme, financed from the European Regional Development Fund (ERDF). The author was an external expert of REURIS project.

orientated towards developing the best revitalisation methods of urban riparian spaces as well as the initiative called RESTORE, aimed to promote good practices in rivers regeneration in Europe[231] (Januchta-Szostak 2017c).

The results of the research programmes allowed to verify the formerly prevailing approach to cultivating the relations river – city, city – catchment area and catchment area – river, giving foundations for the concepts of **integrated water and space management as well as regenerative development** (Cole 2012), which would not only make it possible to minimise the damage but also reconstruct the natural environment of cities and create vital ecosystems, whose integral part consists of human habitats.

Water governance in urban catchment areas

Considering the role of water in shaping urban space involves not only reducing threats of its shortage (droughts), overflow (torrential rains, floods) or its low quality (sewage management, surface and groundwater pollution). It is equally vital that its potential within the scope of maintenance, supply, regulation and cultural services should be used (Kronenberg & Bergier 2010) along with restoration of natural hydrological cycle in the urban catchment areas (Marsalek et al 2008). In the case of increased demand for water and its limited resources, circular economy and renewal of water resources are of key importance. **The Catchment Based Approach** (CaBA)[232], used e.g. in Rotterdam, London or Singapore (cf. Chap. 4), consists in considering the extent of the catchment areas as well as hydraulic and environmental consequences of planning decisions which directly affect flood risk increase. It also requires reverse processes such as determining the capacity of urban areas and the needs of ecosystems in urban strategies and water governance plans and subsequently respecting the guidelines at the stage of establishing spatial development plans and housing conditions.

Regrettably, in most cities the catchment area division is not based on the extent of hydrographical basins of urban watercourses but rather on the zones of combined sewer systems. Consequently, small rivers and streams lose their

---

231 Initiative co-financed by LIFE European Commission funds in close cooperation with ECRR (European Centre for River Restoration).
232 The Catchment Based Approach (CaBA) is a community-led approach that engages people and groups from across society to help improve water environments. <https://catchmentbasedapproach.org/> [accessed: 24.02.2020]. In UK CaBA policy framework was launched in 2013 by Defra. <https://assets.publishing.service.gov.uk/government/uploads/system/uploads/attachment_data/file/204231/pb13934-water-environment-catchment-based-approach.pdf> [accessed: 24.02.2020]

supply resources unless they have already been included into underground sewage system. Sealing and effective draining of cities deteriorates vegetation conditions which weakens evapotranspiration capabilities and ecosystems vitality. Urban water cycle is open: extraction-use-sewage while the precipitation supplies are wasted in the cycle: "from rain to drain", cumulating water problems in urbanised areas.

The necessity of water governance integration with town planning and architecture is emphasised in *The Baltimore Charter for Sustainable Water Systems* (WERF 2007)[233], which imposed an obligation to develop new decentralised urban water systems following natural cycles and formulated the key principles of urban eco-systems engineering (Januchta-Szostak 2011b). Its main postulates include:

- **Onsite and neighborhood treatment** - small-scale technologies that mimic natural membranes and filters and that utilize soils and smart localized controls;
- **Onsite and neighborhood reuse** - closed-loop water systems in residential and commercial buildings, where stormwater and wastewater are treated and reused for landscape irrigation, toilet flushing and cooling, and where minimal waste leaves the site;
- **Green infrastructure** - Rain gardens that trap stormwater and sustain trees and plants. These plants restore beauty and improve air quality in cities, moderate energy flows, and provide potential food sources;
- **Smart Growth** - Patterns of neighborhood development that interconnect nature and the built environment, preserve open space and respect natural drainage flows;
- **Green Cities** - Restoration of natural cycles of water infiltration and evaporation in cities and towns, through localized treatment and groundwater recharge, trees, parks and roof gardens, and stream daylighting and restoration;
- **Watershed restoration** - Restoration of natural watershed flows and functions, through localized water use and recycling into natural wetlands, groundwater, and air. These systems will restore and preserve vegetation and wildlife, and minimize climate changes and warming.

---

233 The Baltimore Charter on Sustainable Water Systems was drafted as a commitment to design new water systems that mimic and work with nature. These systems will both protect public health and safety and will restore natural and human landscapes.

Blue-green infrastructure

Urban built environment is separated from the ground by surface sealing and consequently, vegetation covers tiny, isolated areas. Watercourses regeneration requires restoring water circulation in hydrological cycle (with access to the ground and ecosystems) also in the densely urbanised basins which entails the necessity of a radical change in the approach to shaping architecture and infrastructure. Greenery must become an integral part of both hydrographical and built structures[234] in the tough conditions of competition for space. It is crucial then that "green infrastructure" (GI)[235] and "green architecture" should be popularised, enabling increase in biologically active surface and vegetation cover in cities as well as use of natural processes and ecosystem services in water governance in urban catchment areas. Currently the green infrastructure[236] is planned and used in order to mitigate the consequences of climate changes, absorb carbon dioxide, enhance air and water quality along with biodiversity and biological resilience as well as to promote healthy lifestyle and encourage active leisure.

Blue-green infrastructure follows the principles of nature protection and natural processes sustainment as well as requires purposely including them into spatial planning and territorial development. It is crucially important not only for city climate improvement, health and life quality enhancement, limiting CO_2 concentration but also for sustainability of water management.

Several planning and designing trends, which appeared in the first decade of the 21st century, emphasised the necessity of "sensitising" to water aspects in urbanism. In Australia, *Water-sensitive urban design* (WSUD) popularised spatial planning

---

234 The processes of vegetating cities in the 20th century (e.g. Vienna, London) also contributed to gradual regeneration of catchment areas and water quality improvement, though it was not their main goal.
235 GI (*Green Infrastructure*): strategically planned network of natural and semi-natural areas with other environmental features designed and managed in order to provide wide variety of ecosystem services. Definition included in an announcement of EC on green infrastructure (Green Infrastructure – Enhancing Europe's Natural Capital 2013, European Commission 2013). BGI (*Blue-Green Infrastructure*) – allowing for water components.
236 Green (blue-green) infrastructure consists of a network of urban forests, parks, open terrains, gardens, rivers and wetlands as well as street trees, green walls and roofs of buildings. Also urban systems of small retention (SuDS) are elements of the blue-green infrastructure.

and engineer design considering urban water cycle, including precipitation, ground waters, sewage management and water supply, orientated towards minimising environmental threats as well as enhancement of aesthetic and leisure advantages (JSCWSC 2009). In the United States, similar objectives were formulated for *Low-Impact Development* (LID), in Great Britain known as *Sustainable Urban Drainage System* (SuDS).

Urban green-blue grids (Pötz and Bleuzé 2016) and alternate pro-ecological solutions of rainwater management systems[237] in urban catchment areas are implemented in different range, forms and scale, ranging from **small-scale** projects (covering buildings, public places or small housing estates[238]) through **medium-scale** solutions for districts or city areas[239], to **whole-city projects** allowing for greenery systems and hydrographical networks in spatial development structure of cities[240] (Januchta-Szostak 2017b). In many cases there is no need for sophisticated technological solutions as raised awareness of designers is sufficient. Implementation of green roofs and walls of the buildings, rainsquares (Fig. 3.53), permeable car park surfaces, absorbing basins, bumps or bio-retention ditches[241] allows to use natural qualities of greenery and ground in water retention, infiltration, cleaning and transpiration processes (Januchta-Szostak 2011b).

---

237 Multi-functional rainwater management systems in city public space (TRIP systems) were described by me i.a. in *Water in Urban Public Space. Model Forms of Rainwater and Surface Water Management* (Januchta-Szostak 2011b).
238 E.g. Water Square Benthemplein in Rotterdam, Holland; Arkadien Asperg housing estate near Stuttgart, Germany; Prisma Nürnberg facility, Germany; UW Library in Warsaw etc.
239 E.g. Potsdamer Platz in Mitte district in Berlin, Germany; Kronsberg housing estate–Hannover, Germany; Augustenborg housing estate – Malmö, Sweden; Tanner Springs Park in Portland, Oregon, the USA; Schanrhauser Park near Stuttgart, Germany; UptownNormal Circle in Normal, Illinois, the USA; Sponge Park in New York, the USA; the Sokołówka valley in Łódź, Poland et al.
240 E.g. green streets system in Portland, Oregon, the USA; "*Rotterdam Waterstad 2035*" strategy for Rotterdam, Holland; *ABC Waters* programme for Singapore; Qunli Storm water Park in Haerbin, Heilongjiang, China et al.
241 In dense development areas "green" and "grey" retention-infiltration infrastructure solutions are combined together (e.g. boxes, absorptive wells, draining pipes and chambers) or pre-purifying ones (e.g. separators, sand separators). More: Słyś 2008.

Fig. 3.53: Green infrastructure for retention and reuse of rainwater – a concept for Rataje district in Poznań developed under the supervision of A. Januchta-Szostak (the authors of the study: Izabela Jęczmyk, Məşədi Məcid Cavadzadə, WAPP 2016)

Green architecture

The idea of "green architecture" is closely related to respect for the environment within the scope of pro-ecological technologies, saving raw materials and energy as well as aesthetic and landscape values. Its roots can be found in the respect period which modern architecture creators have begun to rediscover. Already in 1910, Frank Lloyd Wright laid the foundations for pro-ecological architecture – integrated with the landscape and constructed of local materials. Domination of modernism and "machine aesthetics" in search for directions in architecture had prevailed for a long time but art experiments in the 1960s and 1970s, especially *land art* current, influenced the change in the approach to aesthetics of a building in the landscape and its connections with vegetation and water (Nyka 2013). Vernacular architecture played a vital role owing to its references to local development qualities and materials. Creators engaged in shaping green architecture current are i.a. Emilio Ambasz, Peter Noever, Gustav Peichl, William McDonough, Malcolm Wells, Peter Vetsch, Hans Hollein or Charles Jenks. The contemporary realisations such as *Bosco Verticale* in Milan (project by Stefano Boeri, 2012), *Editt Tower* in Singapore (project by TR Hamzah & Yeang, under construction) or eco-architectural visions by Vincent Callebaut create the perspective of green environment of the 20th century cities.

Formulating LEED standards (*Leadership in Energy and Environmental Design*) in 1994 by the Green Building Council was an important step in the process of establishing standards of green architecture as it determined the criteria for design and construction of ecologically responsible buildings. The fundamental LEED[242] principles include:

1. **Sustainable site development** – allowing for the reuse of existing buildings and preservation of the surrounding environment, incorporation of green roofs, and extensive planting throughout and around buildings.
2. **Water resources protection** – by cleaning and recycling of gray water (previously used), precipitation water retention in buildings as well as monitoring water usage and supplies.
3. **Energy efficiency** – increased by orienting buildings to take full advantage of seasonal changes in the sun's position, by the use of diversified and regionally appropriate energy sources, which may – depending on geographic location – include solar, wind, geothermal, biomass, water, or natural gas.
4. **Building material choice** – priority of local materials, recycled, or renewable and those that require the least energy to manufacture, ideally free from harmful chemicals and made of nonpolluting raw ingredients, durable and recyclable.
5. **Indoor environmental quality** – determined by how the individual feels in a space, involving such features as the sense of control over personal space, ventilation, temperature control, and the use of materials that do not emit toxic gases.

Already at the beginning of the 21st century James Wines remarked (2008, p. 64) that "a green architecture must become a basic constant and not remain a mere superficial trend". Data published in 2011 in *Green Outlook 2011 McGraw-Hill Construction Digest* showed that from 2005 to 2010 the "green building market" had grown six fold (Bernstein 2011), especially within the scope of big commercial investments (non-inhabited). However, the *World Green Building Trends 2018*, published by Dodge Data & Analytics (2018)[243], proves that the green

---

242 Wines J., n.d., "Green architecture", *Encyclopaedia Britannica*, <https://www.britannica.com/art/green-architecture> [accessed: 12.11.2018].
243 source: Dodge Data & Analytics, 2018, World Green Building Trends 2018, <https://www.worldgbc.org/sites/default/files/World%20Green%20Building%20Trends%202018%20SMR%20FINAL%2010-11.pdf> [accessed: 2.02.2019].

building activity not only continues to grow across the globe but is expected to exceed certification activity between 2018 and 2021. Recent research results (Dodge Data 2018) reveal the growing impact of social factors such as creating a sense of community, encouraging sustainable business practices and improving the health and well-being of users. Green roofs or walls in the form of vertical gardens, watered by used grey water or retention rainwater, are becoming a standard of buildings applying for BREEAM, LEED[244] or EU Green Building[245] certificates. According to Mc Graw-Hill Construction, using green roofs in the USA rose in 2009 by 16.1 % owing to pro-ecological policy and investment incentives while governance and recycling of water is becoming the most important aspect of green construction (Bernstein 2011). Ecological buildings make it possible to reduce water extraction by 15 % on average. Also, increase in business benefits, such as employees' health, is clearly noticeable which facilitates green architecture development[246]. Exposing economic profits is a key argument orientated towards investors and decision makers, influencing spatial policy trends and economic incentives. Equally important aspects include image, recreational and educational role of green architecture, its influence on building new aesthetic and environmental sensitivity of both the creators and the users of public space. The challenge of green architecture does not only consist in integration with natural landscape and the use of "green material" or application of the latest technologies, based on natural processes of development, change, adjustment, regeneration, energy and matter circulation in artificial urban environment. The challenge also concerns a change in thinking about shaping architecture. According to James Wines[247] "If architecture is to become truly green, then a revolution of form and content—including radical changes in the entire look of

---

244 BREEAM and LEED are two most world popular ecological certification systems, based on the assessment of many unbiased criteria and technical parameters.
245 GreenBuilding is the European Union's programme to enhance the energy efficiency of non-residential buildings. Building owners, who decide on the implementation of modernisation measures may receive the status "GreenBuilding Partner" if they achieve energy consumption reductions of 25% or more. <https://ec.europa.eu/energy/intelligent/projects/en/projects/greenbuilding> [accessed: 24.02.2020].
246 Increasing productivity in "green offices": drop by 39 % of average sick leave days and by 44 % monthly cost of health care, especially for higher level personnel; work efficiency rise by 42.5 % and employees' engagement by 15 % (Bernstein 2011, based on *Green Outlook* 2011, McGraw-Hill Construction).
247 Wines J., n.d., "Green architecture", *Encyclopaedia Britannica*, <https://www.britannica.com/art/green-architecture> [accessed: 12.11.2018].

architecture—is essential. This can only happen if those involved in the building arts create a fundamentally new language that is more contextually integrative, socially responsive, functionally ethical, and visually germane".

Integration of water, environment and space management

The Water Framework Directive (2000/60/EC) gives a chance to recreate biological balance and regenerate the resources, treating ecosystems as an equal user of water. At a region and city scale it is necessary that a vision should be developed along with negotiating management principles within the scope of possessed natural, financial and knowledge assets as well as planning tools and implementation strategies of water sources management.

The situation of the 21st century cities is determined by three fundamental planning paradigm shifts (IWA 2016a): 1. The resources are limited and cities population is constantly growing which entails the necessity of circular economy, 2. Cities densification is a chance for economic development but a threat to life quality in cities, 3. Planning cities is performed under conditions of uncertainty.

International Water Association – IWA in its document titled *IWA Principles for Water-Wise Cities* (IWA 2016a) emphasised four levels of action:

1. **Regenerative Water Services** - the main goal is to ensure public health and satisfy all current needs while protecting the quality and quantity of water resources for future generations by efficient production and use of water, energy and materials.
2. **Water Sensitive Urban Design (WSUD)** seeks the integration of urban planning with the management, protection and conservation of the total urban water cycle to produce urban environments that are 'sensitive' to water sustainability, resilience and liveability co-benefits.
3. **Basin Connected Cities** - the city is intrinsically connected and dependent on the basin it is part of, and which interacts with neighbouring basins. By proactively taking part in basin management, the city secures water, food and energy resources, reduces flood risk and enhances activities contributing to its economic health.
4. **Water Wise Communities** - this level of action is about people building on their existing capacities to govern and plan; professionals becoming more "water-wise" in their area of expertise, so that they can integrate water across sectors, highlighting the co-benefits of integrated solutions to unlock investments.

In order to achieve synergy effect in economic, social and environmental goals pursuit, institutional integration of water governance in urban environment is necessary. *Integrated Urban Water Management* (IUWM, GWP 2013) is based on combining urban areas development with water governance. It consolidates water supply, sewer system, rainwater and sewage management with spatial planning and economic development (Fig. 3.54). IUWM assumes the change in the approach to city management from sector governance to integrated system, based on agreements and negotiations of strategic goals for different areas of urban economy – combining the water sector with other urban sectors such as ground management, construction, energy, transport, in order to avoid fragmentation and duplication of tasks.

Fig. 3.54: Integrated urban water management IUWM (developed by A. Januchta-Szostak, based on: Bahri 2015, p. 69)

The cooperation-based approach engages all the parties concerned to establish priorities, undertake actions and raise responsibility (GWP 2013). Akissa Bahri (2015) emphasises the role of town planning in coordination of water services and communication between particular sectors, administration levels, local communities and stakeholders.

Spatial planning and management in Poland are based on the local authorities' rights to make decisions about forms of municipality spatial development which do not consider the catchment area approach[248]. There are no

---

248 Spatial planning is subject to administrative division while water governance - the division into basins and water regions. Difficulties in integrating spatial and water

legal frameworks and economic tools which would support integrated, pro-ecological approach to space and water management in the cities. The concepts of integrated water governance are known and popularised (Słyś 2013; Krauze & Wagner 2014; Kowalczak 2015; Hlavinek & Zelenakova 2015; Graf & Pyszny 2016), but the structures of sector divisions are so inflexible that the negotiations for urban solutions within the scope of flood risk management are limited to legal permissions in risk areas (river valleys) and not in the catchment areas where the causes of floods are generated. Integrating urban drainage with flood management (Fig. 3.54) is of key importance to achieve efficiency in reducing threats. However, in Poland rainwater management is combined with sewage discharge and belongs to responsibilities of water supply-sewage sector while coordination with town planning is usually limited to agreements on the layout of collector pipes and network connection conditions[249]. Particular areas of city management are supervised by different entities which do not usually coordinate their actions[250] with the others or do it to a minimal extent. Coordination and negotiation of non-standard solutions, exceeding the binding legal framework depends on the efficiency of information flow and operability of planning entities in particular cities. Integrated planning actions based on IUWM in Polish cities are undertaken as a result of bottom up initiatives, like for instance the idea of Blue-Green Network (BGN) in Łódź or programmes on constructing blue-green infrastructure in Wrocław (e.g. "Grow Green").

In 2013, the Ministry of the Environment prepared a "Strategic adaptation plan for sectors and areas sensitive to climate change ..." (SPA 2020 2013), and in 2019 plans for adaptation to climate change were developed in 44 cities with over 100,000 inhabitants. residents, whose aim was to assess the sensitivity and vulnerability of Polish cities to climate change. These new strategic documents can be an important tool for integrated management of space, environment and

---

management result from different areas covered by planning and different priorities of institutions responsible for them. At the local level in cities and municipalities, where the fundamental decisions concerning space management are actually made, there are no binding planning documents on water governance (Januchta-Szostak 2014b).

249 Polish law lacks regulations enabling decentralisation of rain water governance, obligation to retain it in public space and incentives to reduce draining rainwater from private areas. On the contrary, investors were obligated to drain rainwater to collective sewage systems until quite recently.

250 An infamous, yet common example is modernisation of streets and pavements followed by digging them up in order to extend the underground infrastructure system.

water, but their implementation requires many legal and organizational changes in spatial planning.

In Copenhagen, not only the Climate Adaptation Plan (CAP) was implemented, but also the Cloudburst Management Plan (CMP), introduced in 2011. The CMP sets out infrastructure levels and solutions to collect, delay and discharge rainwater runoff. The city also approved a new development vision called Co-Create Copenhagen, highlighting the role of citizens in co-creating Copenhagen.

The examples of water governance in Rotterdam, London and Singapore, provided in the subsequent chapter, illustrate the benefits of catchment area approach and implementation of IUWM principles.

Integration of planning and city governance within the frame of RGB structures enables creating mechanisms which provide *efficiency* in city functioning, *resilience* and flexibility, adaptability as well as spatial order and desirable *life quality*.

4 Responsible cities – vital rivers

> *The consequences of our actions appear sooner than we manage to escape.*
>
> [Jacques Deval]

4.1 Responsibility as a way of return to rivers

The analysis of the stages of conquest and return to rivers in the history of urban development reveals that fundamental changes took place only under the influence of very strong pressure factors or economic, spatial or social barriers[251] e.g. economic collapse, water or space deficit, epidemiological threats, high flood risk level or social unrest. Water problems increase with time. Solutions to current inconveniences (e.g. sewerage of polluted rivers or sealing of city surfaces) caused much more serious problems for the next generations. Urbanisation of an increasing number of areas caused changes in water cycle conditions in catchment areas and hydrological regime of urban watercourses. Water problems flow downstream, so many cities, especially located in upper river course, did not experience the consequences of basins and valleys transformations.

The cities located by river estuaries, like London and Rotterdam, were most prone to the consequences of floods and pollution therefore I chose them to illustrate the problems as well as the stages and methods of creating integrated water management. The restoration of water purity and biological vitality of the Rhine, the Meuse or the Thames has not been fully accomplished but the scale of water problems has forced changes in water and space management in regions and particular cities. In both cities the post-industrial riparian terrains had to be transformed and both of them have had to face the consequences of climate change, though their location and kind of flood risks require different strategies. The size of the city and its water sources are also of great significance. The prospect of London population growth questions possibilities of sufficient water supply. In order to illustrate the problem and applied solutions, I also presented the example of **Singapore**, an Asian city-state, which struggles with water shortage. The rivers of Singapore are short and

---

251 The threshold theory by Bolesław Malisz explains how strong determination is required to overcome barriers to city development and incur the necessary expenses.

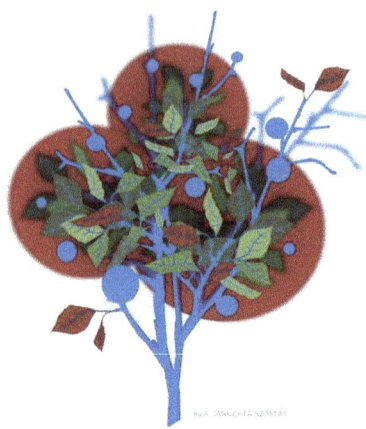

Fig. 4.55: Symbolic RGB structure model of sustainable city based on bionics - *Nature-Based Solutions* (developed by A. Januchta-Szostak) p. 191 (223)

their basins – highly urbanised, therefore the vitality of watercourses depends on rainwater management in the city. The selected cities are similar in one aspect: all of them committed a number of "sins against water". The natural environment had been transformed to such an extent that the consequences of urbanisation began to threaten the stability of urban structures and the inhabitants' existence. The barriers to the development forced a reflection on the approach based on water subjugation and led to changes in the methods of managing rivers and basins.

The return stages and efforts undertaken in numerous cities of Poland, Europe and the world, discussed in Chapter 3, prove that **friendship with rivers requires taking responsibility for water** – responsibility which is not only limited to the areas neighbouring river valleys but also includes tributaries and urban catchment areas. Urban "red crown" (R), consisting of multidimensional structures of Urbs and Civitas, cannot function without purification and retention machine (Fig. 4.55) which comprises water circulation[252] (B) and ecosystems (G). Sustainable rainwater management, circular economy and water management

---

252 Circular water management in cities is part of Circular Economy policy – a regenerative system orientated towards reduction and, ultimately, elimination of waste.

means taking care of eco-hydrographical systems in the entire catchment area and in every building by applying Nature-Based Solutions (NBS)[253].

4.2 Rotterdam – water challenges and assets[254]

Rotterdam developed as a harbour already in the 16th century, owing to its strategic location in the delta of the Rhine and the Meuse (by the New Meuse – delta tributary of the Rhine). The spatial structures of Rotterdam from approx. 1000 AD to contemporary times shown in Fig. 4.56–4.58 illustrate the intensity of the development, especially in the 20th century. At present, the city population amounts to 600,000 inhabitants but it constitutes the southern part of Dutch conurbation of Randstad which is inhabited by approx. 6.7 m people. The population density in the region is highest in Holland (approx. 4,000 people per km²).

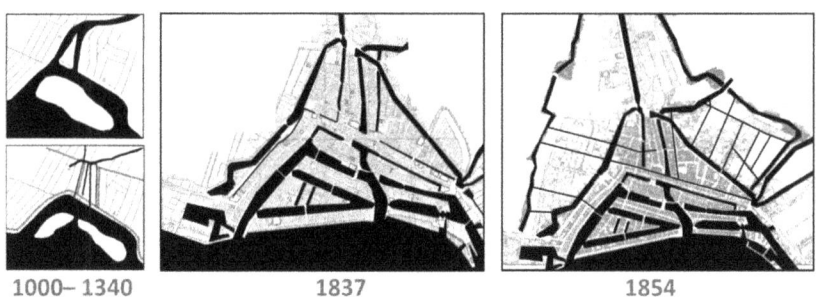

1000–1340   1837   1854

Fig. 4.56: Development of Rotterdam at the beg. of the settlement 1000–1340 (Hooimeijer et al 2005) and industrialisation in the 19th century (developed by A. Januchta-Szostak on the basis of historical plans: A.H. Krap 1837 and W.N. Rose 1854)

---

253 In 2018 the United Nations prepared World Water Development Report (WWDP 2018) and adopted World Water Assessment Programme (WWAP 2018) on Nature-based Solutions.
254 Chapter based on the following articles: Januchta-Szostak 2008, "Kreowanie tożsamości na styku wody i miasta" and Januchta-Szostak 2012, "Shaping cities in the face of the risk of flooding in 21st century. Rotterdam – water city".

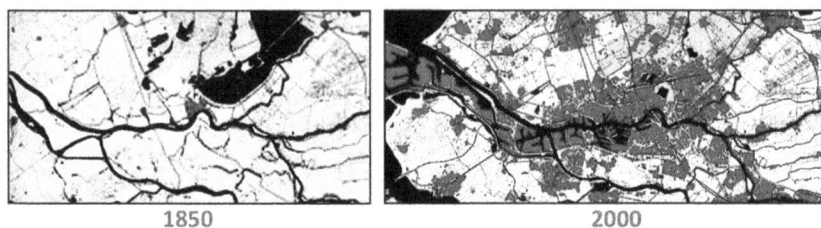

Fig. 4.57: Comparison of urbanisation scale of Rotterdam along with the adjoining towns and harbour structures from the mid-19th century to the 21st century (developed by A. Januchta-Szostak on the basis of: Hooimeijer et al 2005)

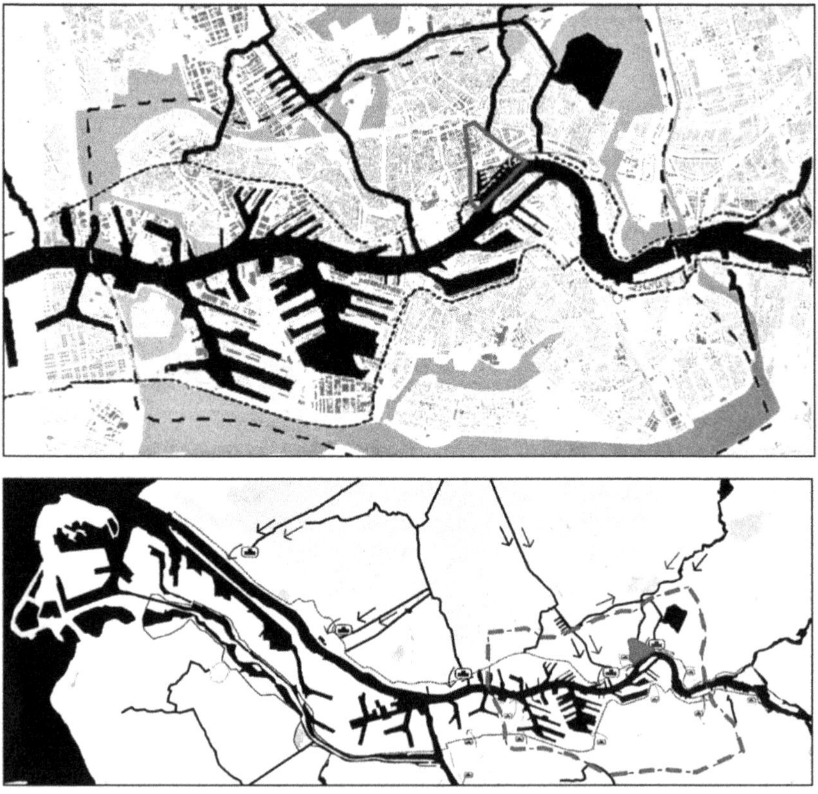

Fig. 4.58: The current view of Rotterdam and its harbour structures. The dashed line marks the modern borders of the city of Rotterdam in the context of the extensive port structures (see figure below) (developed by A. Januchta-Szostak based on of Rotterdam Waterstad... 2005)

Rotterdam is connected with the North Sea by the Nieuwe Waterweg canal (Fig. 4.58) along which harbour terrains of Maasvlakte, Europoort, Pernis, Botlek developed; while the Rhine and the Meuse constitute a part of trans-European waterway, providing excellent access to the industrial areas of France, Germany, Belgium, Switzerland and Austria. Water transport enables shipment of petroleum, metal ore, coal, grain as well as export of petroleum products, chemicals, large-size machines and means of transport. Currently, the harbour in Rotterdam is the biggest one in Europe (once in the world)[255]. Its 40 km port structures are located along the Nieuwe Waterweg canal and the North Sea shore.

Whereas the New Meuse waterfront in the city centre, free from the industrial functions since the 1980s, has been gradually revitalised and adapted to new needs. Nevertheless, the harbour functions determined the identity of Rotterdam along with its functional-spatial structure.

Harbour identity

The trace of relations between the city and its river has been preserved in the riverside urban structures. The specific triangular plan of the old town (Fig. 4.56) resulted from the adjustment to the exchange of goods by waterways. Water City (*Waterstad*) with broad *Boompjes* harbour waterfront was the base of the triangle. Such a shape of the waterfront stemmed from the advantage of integration over defensive factors – the need of creating a wide zone where the city meets the waterway, the place of the exchange of goods and cultural assets as well as prestigious location for the development relating to navigation and freight forwarding.

The advancement of river and oceanic navigation since the 16th century propelled the economy of colonial countries. The war against Spain in the 16th century caused blockade of harbours in Amsterdam and Antwerp which created a development opportunity for Rotterdam. In the 17th century, which was "a golden age" for the Netherlands, the population grew to 50,000. The waterways of the Rhine and the Meuse made the city exceptionally convenient for transport of the goods from overseas colonies to other European countries. The city constituted a multicultural melting pot, sprawling dynamically owing to water transport services, industry, exchange and freight forwarding.

---

255 By the beg. of the 21st century Rotterdam had been the biggest world harbour (record-breaking reloading of 310 m tons in 1973). In 2003 Singapore, and then Shanghai, took over the role of harbour world leaders. Source: Encyklopedia PWN <https://encyklopedia.pwn.pl/haslo/Rotterdam;3969085.html> [accessed: 18.11.2918].

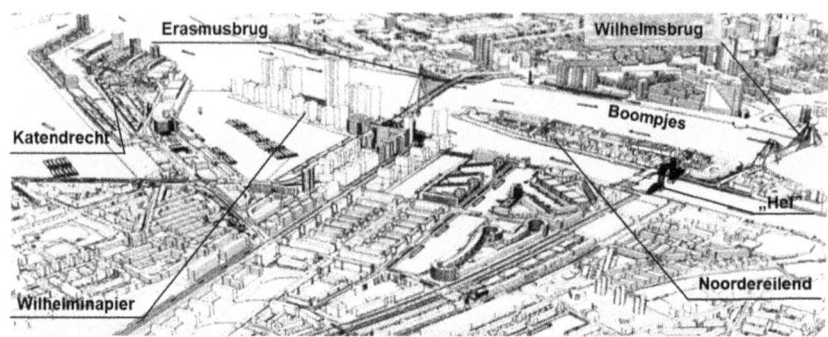

Fig. 4.59: The concept of Kop van Zuid development (project: TeunKoolhaas, 1996). Developed by A. Januchta-Szostak based on: Meyer 1999, p. 355

During the Industrial Revolution, similarly to other European cities, Rotterdam faced the development barrier due to deteriorating sanitary conditions in the city. In 1854 Willem Nicolaas Rose, an urban architect, developed *The Rose Singel Plan* (Fig. 4.56, 1854), which is called today "the first programme for sustainable development". The programme not only allowed for the structure of ownership and development, communication and navigation needs as well as flood precautions but also for landscape relations between water and the city, raising inhabitants' life quality and enhancement of hygienic conditions through water circulation in canals (Waterplan 2... 2007).

A new Rotterdam development programme, formulated in 1870 (author: G.J. de Jongh) was much more pro-industrial. It aimed to expand the city towards the west and the left river bank (Kop van Zuid) as well as to develop further harbour functions which determined both economic and socio-cultural image of the city. In the 19th century, three factors were the catalysts for the city progress 1) the Ruhr area development in Germany, 2) the construction of the Nieuwe Waterweg canal from 1866 to 1872, 3) the resolutions of the Mannheim Convention (1868) permitting navigation on the Rhine.

The first painful experience of identity loss was caused by the destruction of the city due to fascist carpet bombings in 1940, which deprived Rotterdam of historical urban tissue. The inhabitants lost not only their houses but also mental achor points built by characteristic landmarks. Rotterdam became "a city without its heart". After the World War II, the new identity was based on the ethos of work, industrialisation and movement, creating a vision of the city as a European window to the world.

Fig. 4.60: The waterfronts of the Meuse in Rotterdam with the development of Katendrecht, Wilhelminapier and Kop van Zuid in 2007 (photo A. Januchta-Szostak).

A – the Meuse valley, the view of Wilhelminapier and Katendrecht, B - STC - Shipping and Transport College, C – Teatr Luxor, D - Boompjes waterfront and "Manhattan on the Meuse", E – De Brug building

The second step of identity loss followed the relocation of harbour functions. In the 1960s, due to changes in water transport and increase in ship sizes, the main harbour was moved to Europoort. Large transatlantic liners, which had made the waterscape of the city, disappeared. Abandoned docks reminded of cemetery and the lack of harbour life features changed the socio-cultural structure and urban rituals. Apparently, the Euromast, built in 1960 and offering a telescopic view towards the distant ports of Botlek and Europoort over the Nieuwe Waterweg canal, has become an expression of the nostalgia for the irreversible past. In the 1970–80s there was a sharp decline in the attractiveness of the city expressed by a decrease in population, which forced the authorities of Rotterdam to search for new catalysts for development and creating new identity (Meyer 1999).

The city without its monuments due to war damage, filled with degraded infrastructure after relocation of harbours, became a testing ground for town planning and architecture. The *Renewal of Rotterdam* report, 1987, began revitalisation process while Kop van Zuid development project by Teun Koolhaas (1996) created urban planning frame for the new waterfronts of Rotterdam (Fig. 4.59, 4.60). Rediscovery of the Meuse, as the most important benchmark of "collective memory", medium of history and customs as well as representative urban space and a tourist waterway, was the impulse to transplant "the city heart". The New Meuse became the compositional axis of the city, connecting the waterfront public spaces not only with new bridges but also with an architectonical dialogue. While the cultural factors[256] facilitated the image transformation of Rotterdam.

Showcases by leading architects[257] appeared on the Boompjes, Wilhelminapier and Katendrecht waterfronts, creating the landscape of "Manhattan on the Meuse" (Fig. 4.60D). However, as Hans Ibelings (2003)[258] points out, the process is another globalisation trap inevitably leading to erosion of the place identity. The districts of skyscrapers, supposed to symbolise developmental dynamics and potential, emerge in all big world agglomerations, while "Manhattanisation" of cities leads to landscape unification. The building of HAL headquarters[259] (Fig. 4.61) can be an example of the trend, previously being one of the few surviving monuments after the World War II and an essential benchmark in the landscape of the Meuse riverbank. Today, dominated by skyscrapers vicinity, it is a barely noticeable relic of the bygone harbour splendour.

---

256 In 2001 Rotterdam received the title of European Capital of Culture.
257 The main contributors to creation of the architectonic city image were i.a.: Norman Foster, Hans Kollhoff, Mecanoo, OMA Rem Koolhaas, Renzo Piano, UN Studio Van Berkel & Bos, Carel Weeber, Bolles & Wilson, van den Broek&Bakema, KCAP KeesChristiaanse, Fritz van Dongen, EEA Erik van Egeraat et al.
258 Architecture in the Age of Globalization is a subject of a book by Hans Ibelings (2003). The author criticises minimalistic "glass boxes", not matching the place context and aiming only to escalate impressions, which constitute the new landscape of *Zeropolis*, Koolhaas' generic city – a city without identity.
259 Formerly Holland-America Line headquarters (transoceanic passenger transport line) on Wilhelminapier and currently – New York Hotel.

Fig. 4.61: Rotterdam: HAL headquarters (Holland America Line) on Wilhelminapier. On the left: photo from 1959 (by C. Oorthuys, source: Meyer 1999). In the middle: 2007 (photo by A. Januchta-Szostak) – historical HAL building surrounded by new landmarks: Erasmus Bridge, World Port Centre and Montevideo. On the right: photo from 2018 (by 85martijnh)[260]

The concept of connecting Kop van Zuid with the right-bank city centre (project by Teun Koolhaas) has been consistently realised. Dynamic landscape of skyscrapers with distinctive realisations by Rem Koolhaas and OMA is undergoing endless changes (Fig. 4.60–4.62). The Wilhelminapier development is getting denser with recently constructed skyscrapers of New Orleans (project by Álvaro Siza Vieira, 2010) and De Rotterdam (project by OMA, 2013) as well as the Kop van Zuid district with dominating Maastoren – the highest building in Holland (project by Dam & Partners Architecten, 2009, 2018). However, among the galaxy of buildings aspiring to be architectural icons, only the white Erasmus Bridge[261] (project by Ben van Berkel) has truly enhanced Rotterdam's image, becoming a recognisable city landmark (Fig. 4.62).

---

260 Source: <https://pl.tripadvisor.com/Attraction_Review-g188632-d190839-Reviews-Erasmus_Bridge-Rotterdam_South_Holland_Province.html#photos;aggregationId=101&albumid=101&filter=7&ff=327836574> [accessed: 23.11.2018].
261 The bridge was to honour Erasmus from Rotterdam (1467–1536) – one of the greatest humanists of the Renaissance.

Fig. 4.62: Transatlantic and the Erasmus Bridge (project B. Van Berkel) – the old and new symbol of Rotterdam. On the left: photo by A. Januchta-Szostak (2007). On the right: photo by KERSTIN B (2018)[262]

Rotterdam specifically features architectonical experiments like Kubuswoningen (project by Piet Blom, 1984), De Brug (project by JHK Architecten, 2005, Fig. 4.60E), or Floating Pavilion (project by DeltaSync and PublicDomain Architects, 2010), anchored in Rijnhaven harbour. The dynamics of the changing landscape attracts a new generation of creative class which transforms socio-economic structure of the city. Architecture appearing in the former docks (e.g. Lloyd Quarter, project by R. van Dijk, J. Boute, A. de Bruijn) is dedicated to such users. Residential lofts are accessible from the water and from the land and neo-industrial forms of neighbouring buildings open the dialogue with the historical harbour development. STC building (Shipping and Transport College, project by Neutelings Riedijk Architecten bv. 2000–2005), reminding a gigantic periscope directed towards the North Sea (Fig. 4.60A, 4.60B), is the landscape landmark of the quarter. In the 21st century Rotterdam – a multicultural city in transition – treats water as a facilitator for further development and a timeless identity symbol.

Water threats in the time of climate change

Ever since the beginnings of the settlement, the Netherlands has been a testing ground for water town planning, architecture and infrastructure. The space of the country has been so dramatically modified that it is currently entirely dependent on the reliability of the flood protection systems. In the face of the increasing threats resulting from global climate change the conquest approach has changed into coexistence with water. Instead of land reclamation, already at the end of the

---

262 Source: < https://pl.tripadvisor.com/Attraction_Review-g188632-d190839-Reviews-Erasmus_Bridge-Rotterdam_South_Holland_Province.html#photos;aggregationId=101&albumid=101&filter=7&ff=341328117> [accessed: 23.11.2018].

20th century the Dutch began to search for space for water[263]. The embankments in Rotterdam cannot be removed for the simple fact that the major part of it is located below sea level (the lowest location of all Dutch cities) and threatened by sea, river, ground and rainfall waters (Fig. 4.63). "Negotiating the boundary between the city and water" (Nyka 2013) refers both to defending the strategic places of the city and giving space to water.

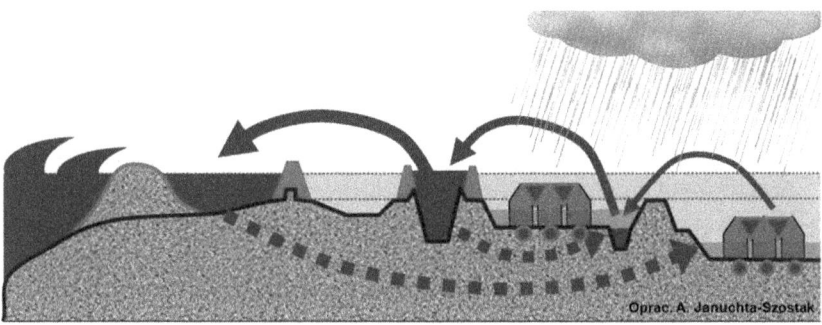

Fig. 4.63: Rotterdam is threatened with sea, river, ground and precipitation waters. The figure illustrates flood threads and the elements of technical infrastructure used to protect and prevent floods (developed by A. Januchta-Szostak)

The latest IPCC (Intergovernmental Panel on Climate Change) reports (IPCC 2018, 2019) predict further sea and ocean level rise (at least 65 cm by the end of the 21st century). The extreme scenario of Atlantis 2035–2070, included aready in the strategy *Rotterdam Waterstad 2035* (2005), predicted sea level rise by 6 m which may place some of the Rotterdam districts even 10 m below sea level. The city located in delta of big European rivers, which flow from beyond the country borders[264], must be prepared for high river overflows, influenced not only by hydro-meteorological factors but also the level of the entire basement urbanisation (e.g. the Ruhr area). Rise in torrential rain frequency and volume as well as in the groundwater level lead to another threat of urban floods. In the areas below sea level, retention of rainfall must be provided and the water gradually pumped to collectors to prevent sea water infiltration (Fig. 4.63, 4.64).

---

263 The direction of changes was shown in the Fifth National Policy Document on Spatial Planning in the Netherlands from 2000 to 2020: *Making space sharing space* (2001) and in *Room for the Rivers* programme (2006).
264 As a consequence of the Rhine floods (1993,1995), also Dutch cities suffered damages (cf. Chapter 2.3) which influenced the approval of *Room for the rivers* programme (2006).

Development strategies for the Water City

Preventing the threats and adaptation to inevitable climatic changes pose one challenge to the city. Another one is the necessity of turning problems into advantages which are the foundations for building resilience, identity and appeal of the Water City. *Rotterdam Waterstad 2035* (2005), a strategy for city development, is the effect of integrated planning and close cooperation between the city and water governance authorities and its main priority directions for **strategic actions include:**

1. **protecting** the city from rising sea level with the use of the "Atlantic Wall", the embankment which is supposed to separate the city from the river and become its main observation boulevard;
2. **creating a specific Rotterdam housing environment on the river**, offering new forms of development rooted in Dutch traditions like "water fortresses" houses on terps, stilts, embankments, piers, pontoon platforms or floating buildings berthed in the former docks[265];
3. **activation of waterways for public transport** (water trams and taxis) and connecting them with land communication system (integrated water-land stations) as well as using numerous canals for extending the system of water routes and increasing retention capacity of the city;
4. **creating public retention spaces**: streets, water squares, plazas and parks on water as well as shaping diverse midtown water environments ranging from urbanised spaces for cultural and sports activities to wild nature enclaves;
5. **concentration of centre-creating functions and city-image creating spaces by the Meuse**: lining the waterfront with spectacular architectural forms, exposed thanks to the water foreground, and activating the riverside public spaces.

The concept of the Atlantic Wall allows for the construction of flood embankments erected 12 m over the existing ground level. Such a barrier can visually separate the city from the river. Looking for advantages, the city authorities pointed out a possibility of creating urban riverside boulevards and exposing vast Rotterdam panorama. Owing to the width of the Meuse river bed, the height of the embankments is hardly visible especially because of the skyscraper architectural forms of Boompjes and Kop van Zuid waterfronts. Unrestrained access to the waterfront is provided thanks to gates with anti-flood

---

265 E.g. Floating Pavilion (project by DeltaSync and PublicDomain Architects, 2010) in Rijnhaven harbour.

bulkheads. The technical flood precautions are integrated with shaping attractive waterside areas for the inhabitants' recreation and integration. Protection from rising sea level also includes actions on regional scale such as revetment of the whole Delta terrain and constructing a strip of artificial islands which constitute a kind of a wave breaker for the urbanised zone of Delta Metropolis[266].

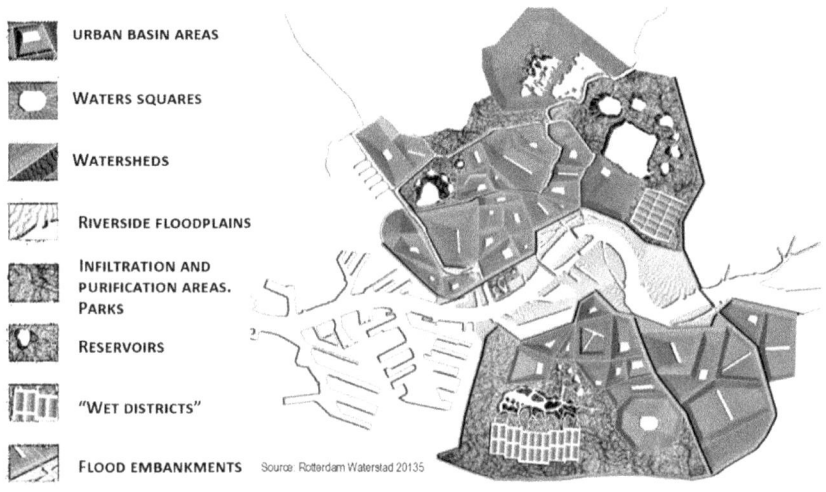

Fig. 4.64: The catchment based approach to rainwater management in Rotterdam. Source: *Rotterdam Waterstad...* 2005, p. 75

Retention of rainwater and its management in the urban catchment areas play an important role in preventing urban floods (Fig. 4.64). High level of ground waters forces rainfall retention on the surface: ranging from green roofs and home gardens, retention systems in a quarter or a district to sustainable water management in the whole city with a system of reduced rate of street drainage and retention reservoirs such as canals, ponds, water squares and parks. Integrated flood risk management, covering the entire catchment area, allows to avoid urban sewer overcharge spill or local floods.

---

266 The zone comprises two conurbations: Dutch Randstad and Belgian VlaamseRuit. The Blue Isles project by West 8 design office, allows for building five islands, from five to 25 km long and 150,000 ha of total surface area. Except for flood protection, they will perform recreational and ecological functions and also constitute a space for extensive development expansion (Kuitert 2008).

A rain in *Rotterdam Waterstad Strategy* is supposed to be a performance and not a trouble, therefore the planned realisations of water square systems include vast range of architectonic forms using water dynamics: floating paths, water mazes, rain cascades etc. (Boer et al. 2010). Water squares not only make it possible to retain rainwater but also change their image and the way they are used depending on seasons and the degree of water filling (Januchta-Szostak 2012). Benthemplein (project by De Urbanisten, 2013) is the first realised water square project which combines an attractive public space with retention capacity (Fig. 4.65).

Fig. 4.65: Benthemplein water square (project by De Urbanisten, 2013). On the left: a bird's-eye view (Rotterdam googlemaps). In the middle and on the right: a view of the square (photo by S. Gajek)

Natural regeneration in the city, so dominated by infrastructure and buildings as Rotterdam, is a considerable challenge but even here the significance of ecosystems and the necessity of recreating biodiversity has been noticed, which resulted in including the blue-green infrastructure into the system of flood protection[267].

Rotterdam Resilience Strategy (2016)[268] exceeds the field of adaptation to climate change and allows for enhancement of resilience to physical, social and economic challenges of the 21st century, establishing six focus areas: 1) Social cohesion and education, 2) Energy transition, 3) Climate adaptation, 4) Cyber use and security, 5) Critical infrastructure, 6) Changing urban governance.

---

267 E.g. through implementing greenery on the roofs, facades and in quarter interiors or the concepts of bio-retention water squares with hydroponic crops.

268 *Rotterdam Resilience Strategy - Ready For the 21st Century* was prepared on the basis of the methodology of *100 Resilient Cities* organisation under patronage of Rockefeller Foundation and approved on 19.05.2016 [accessed: 22.11.2018].

Rotterdam example shows the **"red-blue" way of return**. The city has been implementing *Integrated Sustainable Urban Development Strategies* (ISUDS)[269] which are based on its own experiences of space and water governance within the scope of:

- adjustment to societal challenges,
- transformation of sector management into integrated governance system,
- sustainable integration of economic, social and spatial goals,
- transformation of existing urban areas instead of former external expansion ("from developing green areas to revitalisation of post-industrial areas"),
- local development in connection with general vision,
- change in the role of the city in governance: from regulations to facilitating, from public initiatives to private or public-private undertakings, based on co-creation model.

4.3 London and the Thames – friendship after divorce[270]

London is a city which has committed all the "sins" against water. The short historical outline illustrates the way of expansion – exploitation of the Thames and its tributaries, separating from the catchment area and polluting as well as the way of return – restoring the relation between the river and its urban basin.

Water facilitated development

Already at the beginning of the Common Era, Roman *Londinium* covering 133 ha and inhabited by approx. 30,000 people, became one of the richest cities of the Empire owing to trade contacts with European continent countries (Johnson 1995, p. 30). At the end of the 16th century England entered colonial rivalry and in the 17th century became a naval power. Loading berths, shipyards and stores were developed, changing the riverside area along the Thames into a huge harbour zone. The dense city tissue lacked greenery structures and public spaces[271]. After the Great Fire of London in 1666,

---

269 Based on papers of European Commission workshop, Sevilla 28–29.01. 2016, <http://ec.europa.eu/regional_policy/sources/conferences/udn_seville_2016/rotterdam_south_bank.pdf> [accessed: 17.11.2018].
270 Chapter based on: Januchta-Szostak 2018.
271 Characteristic quality of London greenery structures are city plazas. In the 19th century there were two "lungs of the city": Hide Park, which became a public space already in 1635 and Regent's Park – designed by John Nash and opened for visitors in 1835.

Christopher Wren developed a project of rearranging the spatial structure and opening the city towards the Thames (Fig. 4.66). He propounded implementation of clear system of streets, perpendicular to the Thames, opening the view of the river and joining the city with regulated widened waterfront. Nevertheless, the project was not realised.

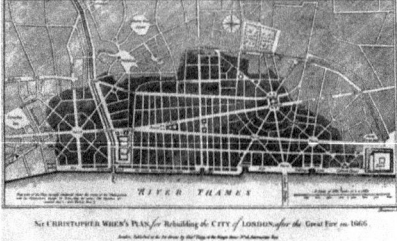

Fig. 4.66: London. On the left: city plan from the second half of the 17th century.[272]. On the right: unrealised plan of London development after the fire in 1666 by Ch. Wren (RIBA Collections)

About 1700 the capital of the British colonial empire was inhabited by nearly 700,000 people. One hundred years later (in 1811) the city population exceeded 4.5 m and the whole agglomeration amounted to 7 m inhabitants. During the Industrial Revolution in the 19th century, seven-fold rise in London population was recorded. Water quality in the Thames began to deteriorate noticeably. The densely developed city, lacking efficient sewage system, suffered from regular outbreaks of dysentery, typhoid fever and cholera which decimated the population[273]. In 1844 under social pressure, the British Parliament approved the construction law which limited the use of septic tanks and imposed the obligation of sewage discharge into the sewer system. Paradoxically, it led to deterioration of the situation and further increase in the river pollution.

---

272 Ogilby and Morgan's Large Scale Map of the City As Rebuilt By 1676 ([s.l.], 1676), British History Online http://www.british-history.ac.uk/no-series/london-map-ogilby-morgan/1676/map [accessed 17 November 2018].

273 Three big cholera epidemics occurred in London between 1831 and 1832 (6,000 victims), 1848–1849 (14,000 victims), 1853–1854 (approx. 10,000 victims). More: Ackroyd 2008, 2011; Clayton 2010; Halliday 2013.

Ecological cost of city development

The Thames was the source of wealth and the symbol of London but it paid a high price for the city's development. In the past, the downstream was wide and shallow with vast marshy areas. For years the riverside swamplands were drained, the waterfronts were heightened and the valley was narrowed by embankments while the river itself – deepened and regulated, having its banks covered with stones and encased in concrete. As a result, the Thames valley in London was narrowed fivefold (from approx. 1.5 km to merely 300 m) and many of its urban tributaries currently flow underground.

Huge ecological disaster which occurred in London in 1858 was called The Great Stink (Halliday 2013). Its direct cause was the Thames contamination with municipal and industrial sewage. Slow river current and its vulnerability to tides led to sedimentation of the sewage on the river bottom instead of washing it to the sea. Record heat wave in 1858 eventuated in water level decrease in the Thames and uncovering the cumulated sewage while the acrid stink paralysed the city life. Thoroughly planned and wide-reaching construction of water supply-sewage system in London, supervised by **Joseph Bazalgette**, eventually produced the intended results. Bazalgette modernised the underground sewage system of London[274], applying pumping stations and big reservoirs where the sewage was treated with the use of chemical-biological processes. Also, building the three Thames waterfronts in Victoria, Chelsea and Vauxhall districts, enhanced the water flow and navigation conditions. According to Peter Ackroyd (2011, p. 80), the significance of the undertaking allows to recognise Bazalgette as one of the biggest London creators like John Nash and Christopher Wren. On the other hand, it is worth mentioning that a number of small rivers (e.g. the Tyburn, the Effra or the Fleet) disappeared from the map of London as a result of the construction of the water supply-sewage discharge system (Fig. 4.67).

Expansion of port-storage and industrial development along the Thames in the 19th century separated the city from the river. The only accessible public spaces for the inhabitants were the midtown river banks. The intensity of municipal and industrial contamination entirely deprived the river of its biocoenosis. Only in the 1990s, as a result of ecological standards change

---

274 Owing to J. Bazalgette's endeavour, 130 km of canals made of bricks and Portland cement were built, which allowed to transport the sewage as far as eastern outskirts of London. All the works were completed form 1869 to 1874. Source: Emily Mann: *Story of cities #14: London's Great Stink heralds a wonder of the industrial world* (English). *The Guardian*, 4th April 2016 [accessed: 2017-08-17].

and the river cleaning campaign, fish appeared in the Thames again. Its urban tributaries had suffered even more. In the 19th and the 20th centuries some of them were transformed into transport channels like the Lee River, others play a role of canalised rainwater collectors (e.g. Ravensbourne River), some urban watercourses are covered and turned into sewage collectors (e.g. the Quaggy), similarly to small streams. London water system includes over 600 km of watercourses. As much as 37 % of the tributaries of the Thames are canalised underground, 25 % have reinforced banks and hydro-technical development and 7 % provide habitat for non-indigenous invasive species (Oates 2012). Due to canalisation and development, London rivers have been separated from the natural environment and their valleys do not constitute the basis for urban green structures.

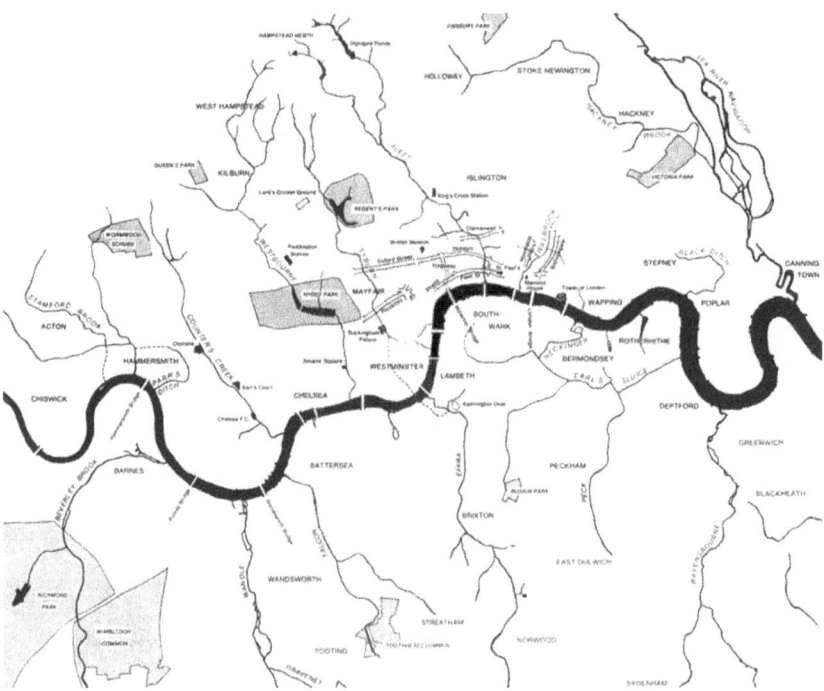

Fig. 4.67: The subterranean rivers of London. Source: The Open Guide to London - london.openguides.org <https://commons.wikimedia.org/wiki/File:London_underground_rivers.jpg> [accessed: 30.07.2018]

Revitalisation of the waterfronts and regeneration of London rivers

The riverside harbour and industrial functions of London had been intensively developing by the end of the 1960s. In the following decade they slowly declined[275] and by the end of the 1970s all the London docks had been abandoned, leaving behind a 21 km² zone of riverside brownfields. Contamination of the Thames and degradation of its waterfront are still substantial even after 40 years of multi-stage revitalisation (Fig. 4.68). St. Katharine Docks (Fig. 4.69) were among the first closed ones in 1968, to be sold and commercially retrofitted at the beginning of the 1970s. Most of the historical warehouse development around the western basin has been demolished and replaced with modern service buildings.

Fig. 4.68: London, view of the Thames and the vast post-industrial areas (photo by A. Januchta-Szostak). On the left: panorama of the city centre with "the O2" Millenium Dome

Fig. 4.69: London, St. Katharine Docks (photo by A. Januchta-Szostak)

---

275 The reason was the introduction of container ship system in transport, requiring bigger ships. It was impossible to adjust the London docks, therefore water transport had to be moved to deeper harbours such as Tilbury and Felixstowe.

In 1971 a governmental report on revitalisation plans was prepared but only ten years later *London Docklands Development Corporation* (LDDC) was founded and a special economic zone was established in order to **revitalise the London Docklands** (Edwards 1992).

The reconstruction of *Canary Wharf* (1980–1990) was quite a challenge (Fig. 4.70) and the experience revealed a number of investment and organisational problems. The standstill was caused not only by the real estate market crisis but also by the lack of communication infrastructure, low quality of the landscape and water pollution. The effectiveness of urban planning depended on coordinated investments in economic-cultural (R), environmental (G) and water (B) sectors. Moreover, the investments were not limited to the Thames and its closest vicinity, as it was originally planned, but had to cover the entire area of the Greater London taking into account the extent of the urban basins. The success conditions for the waterfronts transformation within the scope of **economic-cultural (R - red structures)** field were:

- revitalisation of neighbouring urban areas of *Greenwich* (Fig. 4.71), *Deptford, Southwark - Canada Water* (Fig. 4.72), *Blackwall Basin, Wood Wharf, Royal Docks*;
- enhancement of waterfronts accessibility, raising recreational value and attractiveness of public spaces through creating parks (e.g. *Thames Barrier Park*, project by Allain Provost, Alain Cousseran, 1995–2000, Fig. 4.73), boulevards and walking trails as well as leisure-sports facilities (e.g. *Millennium Dome* "the O2", project by Richard Rogers, 1996, water sports centres, marinas, urban farms e.g. *Mudchute Park and Farm*);
- extension of communication system (integrating *Isle of Dogs* with the London underground system, extending *Jubilee Line* and DLR railway), riverside boulevards, catwalks and bridges (Fig. 4.74, 4.75);
- social-cultural animation (building cultural centres and promotion-education campaigns);
- raising waterfront landscape quality (the Thames – the main axis of city layout – Fig. 4.76, protection of the midtown section landscape, spectacular architectural structures in the waterfront area as well as the city);
- promoting economic, cultural and environmental values of the Thames (and subsequently, also other London rivers e.g. London Rivers Week, organised by Thames21).

Fig. 4.70: London, Canary Wharf (photo by A. Januchta-Szostak). On the left: view from the waterfront towards London city centre panorama, on the right: ongoing process of the extension and modernisation of the docks (2015)

Fig. 4.71: London, ecological housing estates of *Greenwich Village located in the vicinity of Greenwich Ecology Park* (photos by A. Januchta-Szostak)

Within the scope of **environmental changes (G - green structures)**, it appeared necessary to:

- enhance the quality of the environment and riparian green structures as well as pro-ecological development standards (pioneering ecological housing estate of *Greenwich Millennium Village, Greenwich Peninsula Ecology Park* (Fig. 4.71), *Stave Hill Ecological Park)*;
- increase the percentage of greenery areas in the city and their self-regeneration capability as well as mitigate the consequences of urbanisation and climate change (reconstruction of natural assets and ecosystem services through green infrastructure, i.a. water (LWS, Mayor of London 2011) and environmental (LES, Mayor of London 2018) strategies of London;
- connect the scattered green areas into leisure strips (e.g. *the Green Chain Walk*);
- restore biological life of the Thames (controlling contamination runoff, regeneration with the use of ecological engineering tools).

Within the scope of **hydro-technical investment and water governance (B - blue structures):**

- reduction of risk and negative consequences of coastal floods (constructing Thames Barrier – a dam in Woolwich, 1971–1984, Fig. 4.73), river floods

Fig. 4.72: London, Canada Waters (photos by A. Januchta-Szostak). Water purity is the condition for attractiveness of the former docks as the living environment

Fig. 4.73: London, on the left: *Thames Barrier Park*, on the right: Thames Barrier – a big dam protecting the city from storm floods (photos by A. Januchta-Szostak)

(restoring retention capacity of the Thames tributaries – *The London Rivers Action Plan* 2009, as well as *Thames Catchment Flood Management Plan* 2009) and urban floods (implementing SuDS/GI);
- connecting and multi-functional management of London hydrographical system (the *Blue Ribbon Network* policy, 2004);
- sustainable governance of water resources (currently 70 % of drinking water comes from the river and the extraction is increasing while per capita consumption is 150l);

Fig. 4.74: London, new bridges on the Thames – Millennium Bridge (photo by A. Januchta-Szostak)

Fig. 4.75: The Thames – landscape axis of London. Bridges' "gates" visible from the waterway. In the foreground: pedestrian Millennium Bridge (photo by A. Januchta-Szostak)

Fig. 4.76: London, the old and new icons of the Thames waterfront (photo by A. Januchta-Szostak). On the left: Tower Bridge and steep, inaccessible river banks. On the right: panorama of the left bank with the town hall (project by N. Foster) and the Shard skyscraper (project by R. Piano)

- enhancement of water quality of the Thames and its tributaries (*Integrated River Basin Management* IRBM, *River Basin Management Plan*s; *Thames Tideway Tunnel* 2016–2023);
- implementation of the blue-green infrastructure and SuDS (e.g. the residential centre of the Barbican, Fig. 4.77).

182 Responsible cities – vital rivers

Fig. 4.77: London, the Barbican centre (photo by A. Januchta-Szostak, sat photo: Google maps)

Currently, the waterfronts are tourist showcase – the exhibition space for historical city panorama and new spectacular architecture[276]. Nevertheless, transformation of the metropolis into water and inhabitants friendly city still requires considerable effort, bearing in mind the fact that London population is constantly growing and climate changes pose new challenges.

Return strategies

The process of restoring eco-hydrological and socio-cultural vitality to London rivers has just begun and requires multi-stage, reconstructive, systemic undertakings, not only concerning ecological regeneration and retrofitting of water supply-sewage systems but also involving education and society engagement as well as public-private partnership.

Ideas orientated towards limiting spatial expansion of London and improving leisure conditions appeared already at the beginning of the 20th century[277] in the concept of *The Metropolitan Green Belt* surrounding London. Initially, limiting the possibility of development in suburban areas mounted resistance because of radical increase in ground prices which reached 70 % of the new development. *Green Belt* became a space for weekend leisure but growing densification of the inner city structure worsened the conditions for daily recreation. Separate parks and squares, forming greenery-spots layout, did not provide continuity of nature and leisure. As a response, the concept of

---

276 E.g. City complex of skyscrapers, London town hall (project by Sir Norman Foster, 2002) or the Shard (project by Renzo Piano, 2012) – the highest skyscraper in Western Europe.
277 The concept of "Green Belt" appeared already in 1935 and in 1947 was considered in spatial development plans.

the Green Chain Walk[278] was implemented which allowed to connect the green areas into a system of linked parks in the city outskirts. The network of radial corridors connecting the internal and external systems of greenery is based on river valleys (Drapella-Hermansdorfer 2005). One of the first realisations undertaken in 1967 was creating a zone of riverside parks and leisure-sports facilities called *Lee Valley Regional Park* in the post-industrial areas along the Lee river in East London (Bernat 2007, p. 260). The Summer Olympic Games in London in 2012 dramatically changed the image of the Lee River valley. *Queen Elizabeth Olympic Park* covered 2.5 km² including the Olympic Village and a number of spectacular sports facilities[279] as well as a system of paths and public spaces with playgrounds and leisure spots. Equally important was the approach to shaping the river landscape, on the one hand focused on retaining the industrial identity (Fig. 4.78) on the other – re-naturalising sections of the riverbed and creating ecosystems based on indigenous plant species (Fig. 4.79).

Fig. 4.78: Development of the Lee river orientated towards enhancement of waterfront accessibility and cultural identity of the place (photo by A. Januchta-Szostak)

---

278 In 1977 the southern part of the "green belt" was realised, connecting open areas between the Thames and Crystal Palace Park in the system of public greenery.
279 I.a. Aquatics Centre, project by Z. Hadid; or Velodrome, project by Hopkins Architects

Fig. 4.79: Naturalised sections of the Lee river – reed biotopes facilitate water purification processes (photo by A. Januchta-Szostak)

The Blue Ribbon Network (BRN), a strategic element of the Greater London Plan, is a spatial policy including the Thames and the whole integrated London hydrographical system (Fig. 4.80): rivers, their tributaries, canals as well as connected water areas: docks, lakes, artificial water reservoirs. What is important, the idea of the network allows not only for surface waters but also sections of watercourses canalised underground and their interconnections and relations with the surroundings. *The London Plan* (Mayor of London 2004) follows the main principles of development of *The Blue Ribbon Network*:

- **protection and facilitation of multi-functional character of the network** through preference of direct water vicinity functions;
- **public accessibility** as well as sport, recreation and education functions priority within the public space of London which is supposed to provide social integration, revitalisation and development of neglected areas;
- **simulation of economic growth** thanks to regeneration and activation of the waterfronts and extension of London system of canals and rivers;
- **enhancement of communication** and accessibility of London through increase in use of the waterway network to transport passengers and goods as well as development of recreation, water and waterside tourism (the system of pedestrian and cycling trails along the watercourses);
- **protection and enhancement of bio- and landscape** diversity of the network which contributes to vitality and distinctiveness of many parts of London;

London and the Thames – friendship after divorce 185

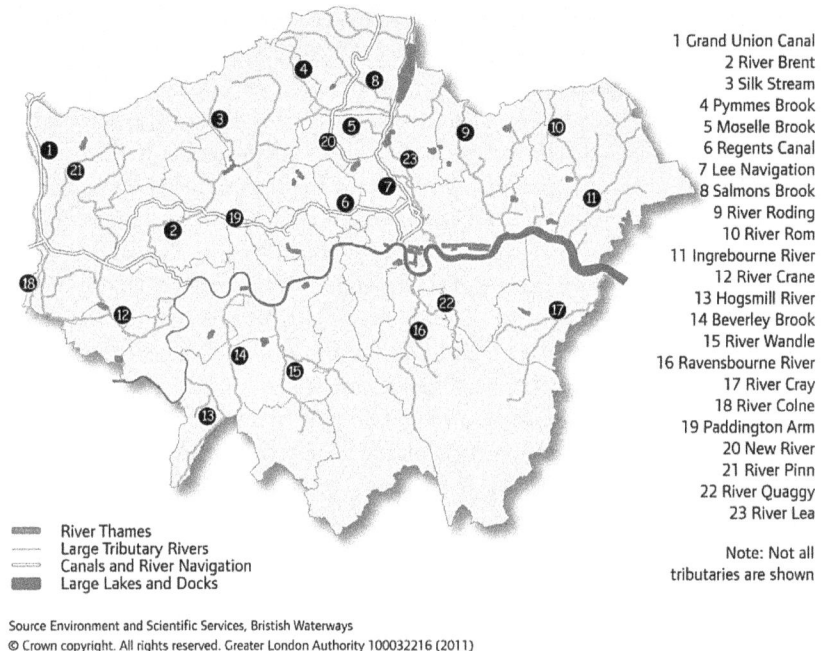

Fig. 4.80: *The Blue Ribbon Network* (BRN) in London, source: <https://www.london.gov.uk/what-we-do/planning/london-plan/current-london-plan/london-plan-chapter-seven-londons-living-space-4> [accessed: 7.09.2018]

- **diversification and sustainability** of water sources as well as sewage treatment methods, using the potential of the network to manage flood risk in connection with necessary adaptation to climate change.

The principles and plans of the city orientated towards rivers regeneration were formulated in an operational plan called *The London Rivers Action Plan* (River Restoration Centre 2009), in agreement with the *Environment Agency* and a number of ecological organisations[280].

The Greater London surface area (1,579 km²) covers a considerable part of the Thames basin whereas 80 % of its flood plains are urbanised. Approximately 45,000

---

280 The signatories of the plan were the Mayor of London and representatives of the *Environment Agency, Natural England, Thames Rivers Restoration Trust, London Wildlife Trust, WWF-UK, The River Restoration Centre*. The city coordinates cooperation of different organisations to achieve common aims.

properties are threatened with 100-year flood (1 % probability of occurrence) majority of which are "socially deprived areas" (River Restoration Centre 2009). Nearly 40 % of London area has impermeable surface therefore torrential rains cause immediate overflows in watercourses and sewage systems spills every season whereas flood warning time is extremely short. Adjusting the development to flood threats and a river restoration programme, promoted by the Environment Agency within the scope of the *Thames Catchment Flood Management Plan* (Environment Agency 2009), are supposed to be the fundamental means of flood risk reduction in London.

In 2011 Boris Johnson, then Mayor of London, emphasised that "We should undo the hydrophobic policies of the 1960s – which saw natural rivers and waterways encased in concrete – and find ways to work in harmony with water in our landscape to ease the consequences of heavy rainfall and beautify our city at the same time. These strategies will help us to stand on the shoulders of Bazalgette and future-proof London for the challenges ahead" (*London Water Strategy*, Mayor of London 2011). The strategy focused not only on providing the city with water supplies and protection against floods but also on considering the future of water in London. Its fundamental element consisted in the use of green infrastructure and more creative approach to managing surface water flood risk, orientated towards slowing the rate of runoff "from rain to drain" as well as economic use of rainwater in order to reduce the demand for treated water.

The significance of green infrastructure[281] and SuDS in the flood risk management as well as enhancement of London waters purity (Proposal 8.2) was even more emphasised in the latest *London Environment Strategy* (Mayor of London 2018). Sustainable drainage systems are also the strategic element influencing urban vegetation increase. They allow to improve vegetative conditions (supplying ground waters), reduce urban areas sealing and popularise the use of green infrastructure in the processes of retention, infiltration and rainwater management as well as enhancement of landscape and recreation advantages of the city. The strategy allows for ecosystem services rendered by London natural assets (open terrains, air, water, wild flora and fauna)[282] such as clean air and water which benefit Londoners' welfare and city economy.

---

281 The green infrastructure is planned and used in order to promote healthier lifestyle, reduce consequences of climate change, improve the air and water quality, encourage active leisure time, absorb carbon dioxide as well as enhance biodiversity and ecological resilience (Mayor of London 2018, p. 135).

282 There are 8 m trees in London which generate profit of 133 m pounds yearly within the scope of air pollution removal, carbon dioxide absorption and urban floods prevention (Mayor of London 2018, pp. 137–138).

Extension of the blue-green infrastructure is closely connected with natural regeneration[283] and re-naturalisation of London rivers and its aim is to restore biodiversity of local ecosystems and environmental balance in order to recreate its capacity to **self-regenerate**. The goal can be only achieved through cooperation of the authorities, the inhabitants and numerous governmental and non-governmental organisations[284] but consistent implementation of the long-term strategy is undoubtedly effective. Owing to information accessibility, multiple guides and educational campaigns (e.g. *London Rivers Week* Thames21) increasingly more local bottom-up initiatives are undertaken in different London districts. Since the publication of *London Rivers Action Plan* (River Restoration Centre 2009) by 2018, 27 km of rivers had been re-naturalised or uncovered in London[285], i.a. the Brent, the Crane, the Colne, the Roding, the Beam, the Ingrebourne and the Lee. The restoration of swamplands and ecosystems of London rivers not only provides space for water (B) and enhances biodiversity (G) but also rises inhabitants' life quality and their ecological awareness (R) which gradually transforms London into water, environment and people friendly city.

4.4 Singapore – "water wise city"

Water "tiger"

Singapore – a city-state and an island of 572 km^2 surface area has nearly 5.5 m inhabitants (2012) and is one of the most densely populated world countries (approx. 7,800 people/km^2). Most of the country surface area (90 %) consists of urbanised and industrial areas which are constantly expanding because of ongoing extension of harbour-lined coast of the island, while the natural areas cover merely 4.5 % and are protected in sanctuaries (Fig. 4.81). In the 19th century the island was a British colony while the trade harbour founded by the

---

283 Natural regeneration covers a range of diverse actions: from local ones such as replacing concrete casing with natural materials and vegetation, recreating flora and fauna habitats, removing invasive species to regional undertakings, covering transformations of whole valleys and flood plains, removing embankments and revetments, reconstructing river meanders and recovery of natural erosion-sedimentation processes (*Rivers by Design...* 2013).
284 *London Rivers Restoration Group* comprises *Environment Agency, Thames21, the Greater London Authority, the Wandle Trust, London Wildlife Trust, Green Corridor, the Thames Estuary Partnership and the River Restoration Centre*, and it belongs to a bigger organisation called *the Catchment Partnerships in London* (CPiL).
285 More: *London Rivers Action Plan*, 2009; Thames21: London Rivers Week 2018 <https://www.thames21.org.uk/joinacampaign/londonriversweek/> [accessed: 24.08.2018].

British East India Company in 1819 was used as a marine corps base. After joining the Federation of Malaya in 1963, Singapore gained independence in 1965 and in a span of nearly 50 years transformed itself from a developing to a highly developed country[286].

Fig. 4.81: The current (2018) spatial development plan of Singapore (Urban Redevelopment Authority, <https://www.ura.gov.sg/maps2/?service=mp> [accessed: 18.11.2018])

Singapore, one of "Asian tigers", owes its economic success to not only the convenient location and political stability but also the system of planning and managing the economy (e.g. encouraging investment) as well as constant enhancement of transport and industry. The *Agency for Science Technology and Research* (A*STAR) – a national institution, facilitating development of high technology industry[287] – plays an important role in this sector.

---

286 In terms of GDP Singapore is one of the world leaders. In 2007 economic growth amounted to 7.7 %. The economy is mostly based on services (68.7 % GDP), industry and construction. Singapore is the biggest financial, banking and trade centre in South-East Asia. Source: *Encyklopedia PWN*/Singapur – Gospodarka, <http://encyklopedia.pwn.pl/haslo/Singapur-Gospodarka;4575324.html> [accessed: 18.11.2018].
287 Source: Encyklopedia PWN/Singapur – Gospodarka, <http://encyklopedia.pwn.pl/haslo/Singapur-Gospodarka;4575324.html> [accessed: 18.11.2018].

The harbour in Singapore is one of largest in the world in terms of tranship‑ment (like Shanghai, Rotterdam and Hong Kong). Deficit of space and local natural resources makes the shipment of food and raw materials indispensable, therefore the harbour in Singapore is not only a catalyst for economy but also an economic necessity.

Unremitting development of harbour functions has determined the land‑scape of the coastline (Fig. 4.82) especially on the south-west outskirts of the island but the terminals in the city area are supposed to cease operating by 2040. After closing them, Tuas Megaport[288] will be the only harbour in Singapore and "the single largest fully-automated terminal in the world", according to Nelson Quek, Head of Tuas Development – PSA Corporation Limited. The new harbour is supposed to be twice as big as the new city of Ang Mo Kio. Its 8.6 km long zone of infrastructure will be capable of serving the biggest container ships in the world[289]. Exclusive yacht ports and passenger terminals, such as *Marina South Pier,* are located on the midtown waterfront of the Central Region. Representative waterfronts and spectacular architecture (*Marina Bay Sands*, *Art Science Museum*, and *Theatres on the Bay*, Fig. 4.83) dominate in the heart of the city situated around Marina Bay (Fig. 4.84). In 2012 the Gardens by the Bay (Fig. 4.85) became a showcase of the city as a symbol of the aim of Singapore which is transforming itself into a **City of Gardens and Water.**

Fig. 4.82: Keppel Container Terminal in Singapore (photo by Kroisenbrunner, <https://en.wikipedia.org/wiki/Port_of_Singapore> accessed: [18.11.2018])

---

288 Heng, Daniel (7 February 2018). "Why Singapore needs Tuas mega port to keep ruling the seas". Channel News Asia. Retrieved 12 March 2018. Source: https://en.wikipedia.org/wiki/Port_of_Singapore#cite_note-Heng-11> [accessed: 18.11.2018].
289 Op. cit.

Fig. 4.83: Singapore, on the left: Marina Bay Sands and Art Science Museum, on the right: Esplanade – Theatres on the Bay (photo by: Marcin Konsek, <https://pl.wikipedia.org/wiki/Singapur> [accessed: 18.11.2018])

Fig. 4.84: Singapore, the view of the Marina Bay in the city centre (photo by: chensiyuan, 2012, https://en.wikipedia.org/wiki/Marina_Bay,_Singapore#/media/File:1_singapore_flyer_view_2012.jpg> [accessed: 18.11.2018])

Water deficit

Rapid population growth and water deficit forced cities like Singapore to implement new water governance strategies, closely integrated with spatial and town planning and architecture. Formerly, the island was covered with humid rainforests and coastal mangrove bushes. Regrettably, they were cleared due to the radical development of the city-country. For many years the scarce water resources had been wasted due to fast discharge to the sea through the canalised rivers. The problem was also high level of pollution causing nature degradation.

Currently, Singapore can boast the biggest forest area percentage in the world, which amounts almost to 29.3 %[290], and integration of water governance with environment regeneration.

Singapore authorities prioritised diversification of water supply which is sourced from local basins (including rainwater from urbanised areas), import from Malaya, desalinated sea water and treated sewage (so called NEWater – high quality reclaimed water). Biogas produced in the process of sewage treatment is additional source of energy (data *Public Utilities Board*). Use of drinking water is also limited in households (currently 155 dm^3 per twenty-four hours per inhabitant).

The strategy of city-basin management

The young city-state had the advantageous opportunity to originally create integrated institutional structures which facilitate planning and management of the space and island resources. The first Concept Development Plan, formulated in 1971, is a strategic document and a long-term vision of spatial development and transport which determines the city development directions for the following 40–50 years. Regular updating every ten years requires close cooperation of all governmental agencies in order to integrate the possibilities of regeneration of natural island assets with meeting the needs of the inhabitants and the economy. The transition to a circular economy requires integrated urban resource management. Public utilities play a special role, as they are responsible for the flow of resources in and outside the city, as well as the relationship with the water and waste cycle (IWA 2016b).

Singapore's National Water Agency called *Public Utilities Board* (PUB)[291] belongs to the Ministry of Environment and Water Resources responsible for water supply, management of the catchment area and waste water in the city with the use of circular economy. PUB closely cooperates with NEA (*National Environment Agency*)[292] in order to minimise the environmental impact as well

---

290 Interactive Web tool Treepedia, developed by researchers from MIT Senseable City Lab in association with World Economic Forum, facilitates interactive maps of the biggest world cities along with information about the percentage of greenery, based on data from Google Street View. Only Tampa in Florida (36.1 %) has a higher rate than Singapore, while London has merely 12.7 % of surface area covered with trees. Source: <http://senseable.mit.edu/treepedia> [accessed: 18.11.2018].
291 Public Utilities Board, <https://www.pub.gov.sg/> [accessed: 18.11.2018].
292 National Environment Agency, <https://www.nea.gov.sg/> [accessed: 23.11.2018]. NEA estimates the consequences of contamination resulting from all planned investments.

as with URA (*Urban Redevelopment Authority*)[293] responsible for architecture, spatial and town planning. The very name of URA includes the mission and the goals of actions which are orientated towards regeneration and revitalisation of already urbanised structures. The change in actions direction from external expansion to enhancing effectiveness of management and reuse of inner resources of the city, resulting from limited island space, leads to search for innovative solutions and close cooperation with the research sector[294].

Due to monsoon rains and tight sealing of catchment area surface, Singapore is prone to urban floods. To enhance flood protection, PUB adopted a *Source-Pathway-Receptor* (SPR) approach, which enhances flexibility and capability of adaptation to extreme precipitation in the whole city drainage system. The analysis of the basin allowed for determining the areas which generate the biggest surface runoff ("a source") and the areas where floods are likely to occur ("a receptor") as well as modifying a system of links between them ("a pathway").

Integrated stormwater and flood risk management based on SPR includes the following actions (PUB 2014):

- **at "a source"**, orientated towards reducing the runoff rate, retention and water governance within buildings, quarters and districts using grey infrastructure to store, pre-purify and re-circulate grey and rain water as well as green infrastructure in the shape of green roofs and walls, reservoirs and bio-retention channels, rain gardens etc., making it possible to biologically clean the urban space and increase its social attractiveness;
- **on "a pathway"**, orientated towards enhancement of flexibility of the source-receptor connection (e.g. through constructing canals between basins) as well as retention capacity of the flow through broadening and deepening canals and constructing retention reservoirs. Re-naturalisation of rivers and streams also contributes to increasing biodiversity and use of ecosystems in biological purification of surface water during their flow through urban catchment areas;
- in **"a receptor"**, areas especially threatened with urban floods, where investments aim not only to determine the terrains of controlled floods and raise buildings

---

293 Urban Redevelopment Authority, <https://www.ura.gov.sg/Corporate/> [accessed: 18.11.2018].

294 In Singapore there are six autonomous universities and specialised training institutions like *Singapore Water Academy* (SgWA), founded by PUB, which offers programmes for integrated urban management professionals and has been managing *The Singapore International Water Week* (SIWW), a global platform to exchange water innovative solutions, for ten years.

resilience (protecting basements, high ground floors, anti-flood dams etc.) but also streamline the systems of monitoring, warning and inhabitants' evacuation.

Singapore authorities prioritised new technologies, integration of spatial planning with water governance as well as participation of the society and business in water resources protection (*3P Approach*). One of the applications of the strategy in the process of protection and responsible use of water sources and the environment is a programme called *Water for All: Conserve, Value, Enjoy*. Urban catchment areas currently cover two-thirds of the country's surface area, therefore enhancing the urban environment quality is not only a condition for clean water supply but also for satisfactory quality of life for the inhabitants. Driven by the vision of sparkling rivers with landscaped banks, kayakers paddling leisurely in the streams with clean waterways flowing into the picturesque lakes, Singapore has undertaken the challenge of transforming itself into a "**City of Gardens and Water**" (PUB)[295]. The state water system consists of nearly 8,000 km of waterways[296] and 17 reservoirs. In order to use its potential in 2006, PUB implemented **ABC Waters** programme (*Active, Beautiful, Clean Waters*, 2007–2030) and management strategy, which aims to (PUB 2018):

A. (Active) activate waterside public spaces, attract people to water, develop a sense of co-responsibility for the environment and water resources;
B. (Beautiful) beautify water reservoirs and green areas in connection with urban landscape; manage to achieve not only anti-flood and retention goals but also to enhance the landscape aesthetics and create tourist and urban water attractions;
C. (Clean) improve water quality through recovering connections with basin and ecosystems, extending blue-green infrastructure, educating ecologically as well as developing attachments between the inhabitants and the water.

PUB is the institution responsible for the realisation of the programme but the actions are performed in cooperation with URA and include i.a. implementation of SuDS systems, restoration of rivers and streams, construction of blue-green networks and retention reservoirs with the use of green infrastructure. The programme realisation included developing the water management plan integrated with the space development plan (with over 100 prospective implementation localisations) as well as creating legal framework and project guidelines within the scope of shaping green infrastructure and architecture. The funds for

---

295 Active, Beautiful, Clean Waters Programme, <https://www.pub.gov.sg/abcwaters/about> [accessed: 18.11.2018].
296 Rivers in Singapore are short. The longest of them, the Seletar, is merely 16 km long.

realisation of the programme are obtained from i.a. taxes: *Water Conservation Tax* – a tax for water sources recovery, *Waterborne Fee* – a tax calculated on the used water volume, *Sanitary Appliance Fee* – a charge for collecting municipal sewage based on the number of sanitary appliances.

The strategy of sustainable urban water management of Singapore, based on the principles of "Water-Wise City" (WWC) and consists in (IWA n.d.):

- **vision** of "*Singapore: City of Gardens and Water*", ensuring an efficient, adequate and sustainable supply of water;
- **integrated governance**: PUB (within the frame of the Ministry of Environment and Water Resources) is responsible for all the aspects of water management, whereas URA, as the central planning authority that integrates urban planning across the various sectors of water, waste, transport, architecture and landscaping;
- **knowledge & capacity**: the scientific background is provided by the Singapore Water Academy (SgWA) whereas the *Singapore International Water Week* is a global platform to share and co-create innovative water solutions. Innovativeness is financially supported by the *National Research Foundation*;
- **planning tools** such as Water Master Plan - the long-term water management plan integrated with URA's Master Plan – spatial development plan;
- **implementation tools** - pricing water to recover both the full cost of its production and supply, investments in the future, through research funding from the National Research Foundation. Within the scope of *People-Private-Public* (PPP) approach, the implemented regulations and programmes aim to encourage to save water and develop the inhabitants' co-responsibility for the landscape and water resources.

Green infrastructure and architecture

Systems of blue-green infrastructure (BGI), implemented within the whole city but also within every district and building, serve retention and regaining water, minimisation of flood risk as well as enhancement of the environment quality and inhabitants' life conditions. Numerous programmes (e.g. *ABC Waters, Friends of Water, Climate Action SG,* or *Eco-Friend Awards*) make it possible to shape the development and landscape of Singapore allowing for the whole catchment area. Anti-flood investments, based on SPR principles (source-pathway-receptor), are closely related to creating blue-green networks of riverside parks and retention-leisure reservoirs. While the built environment corresponds to the vision of a city orientated towards increasing the surface of greenery and water. Architectonic-construction regulations are complemented with the Code of

Practice on Surface Water Drainage and the project guidelines (PUB 2018). The development of BGI is supported by diverse offer of educational programmes, grants and prizes, prepared by PUB and NEA[297]. *Gardens by the Bay* project[298] (Fig. 4.85) is one of the most spectacular examples of numerous investments realised within the *ABC Waters*[299] programme.

A Phot: dronepicr, 2017. Supertree Grove Singapore

B Phot. Jan, Skyway @ Gardens By The Bay, 2014
https://www.flickr.com/photos/jhecking/14682030600/

C Phot: RudolfSimon, 2012. GbB — Conservatories

D Phot: Allie Caulfield. Flickr: 2012. The Cloud Mountain

Fig. 4.85: Singapore, Bay South Garden (photo: *CC*, <https://en.wikipedia.org/wiki/Gardens_by_the_Bay> [accessed: 18.11.2018])

---

297 3P Partnership Fund is a grant administered by National Environment Agency (NEA), meant for organisations and firms to encourage them to cooperate in developing innovative and sustainable environmental solutions promoting responsibility for the environment in local community.

298 In 2011 the first part of the Bay East Garden was opened to the public, which ultimately is supposed to cover 32 ha area with a two-km promenade over the Marina Reservoir. Bay Central Garden (surface 15 ha) will be a link between Bay South and Bay East Gardens with a three-km promenade. The project won 16 prizes, including *The Landscape Institute Awards* 2013 for adaptation to climate change, *World Building of the Year* 2012 reward as well as *IFLA Cultural and Urban Landscape Award of Excellence*.

299 The list of projects, case studies and project guidelines is included in: PUB, 2018, *Active, Beautiful, Clean Waters. Design Guidelines*. 4th Edition.

Bay South Garden (project by Wilkinson Eyre Architects), 54 ha surface, opened to the public in 2012, is the first and the largest of the three designed gardens in *Gardens by the Bay* area in Singapore. Distinctive forms of the "Supertrees" (Fig. 4.85B) became Singapore symbol illustrating the direction for the city development whereas the complex of orangeries (Fig. 4.85C), located just by the Marina Bay, reminds two giant waves, offering spectacular spatial impressions both outside and inside the building (Fig. 4.85D).

Glass structures of cooled winter gardens: *Flower Dome*[300] and *Cloud Forest* (Fig. 4.86) are not only landscape showcases but also promote sustainable and energy-saving technological systems, allowing to minimise ecological footprint. Rainwater is collected from the surface of the buildings and circulates in the cooling system. The system is connected with the structures of the "Supertrees" which serve both to exhaust hot air and cool circulating water (Fig. 4.86). Rainwater runoff from the paved surfaces, constituting 35 % of the area, is cleaned with the help of reed biotopes and greensand filter before being channelled to Marina reservoir. Purified water is used to water vertical greenery on the Supertrees structures. Both sky and land educational paths as well as interactive information panels serve education through leisure, raising the recipients' knowledge on the significance of biodiversity, ecosystems services and principles of purifying water (PUB 2018).

In Singapore not only separate ecological investments are undertaken, the city is also reconstructing blue-green networks through re-naturalisation of rivers and building eco-bridges. The Kallang river (10 km long) drains the basin area of approx. 600 ha. Sections of the canalised river are being re-naturalised and transformed into linear parks within the framework of the *ABC Water* programme. Realisation of the **Kallang River Bishan-Ang Mo Kio Park** (project by Atelier Dreiseitl, 62 ha, 2012) brought spectacular effects resulting from integration of hydro-graphic network and natural structures with the system of public spaces. Transforming the concrete riverbed of the Kallang river (2.7 km long) into a meandering stream, increased the length of the watercourse and retention capacity of the valley which allowed to manage the monsoon rainfall and prevent floods (Fig. 4.87). Diversifying riverside ecosystems and boosting littoral zone enhanced self-purification capacity of the watercourse while the Bishan Park area has become a habitat for numerous plant and animal species and an appealing place of leisure and ecological education as well.

---

300 Flower Dome is the largest in the world glass structure without supporting posts.

Singapore – "water wise city" 197

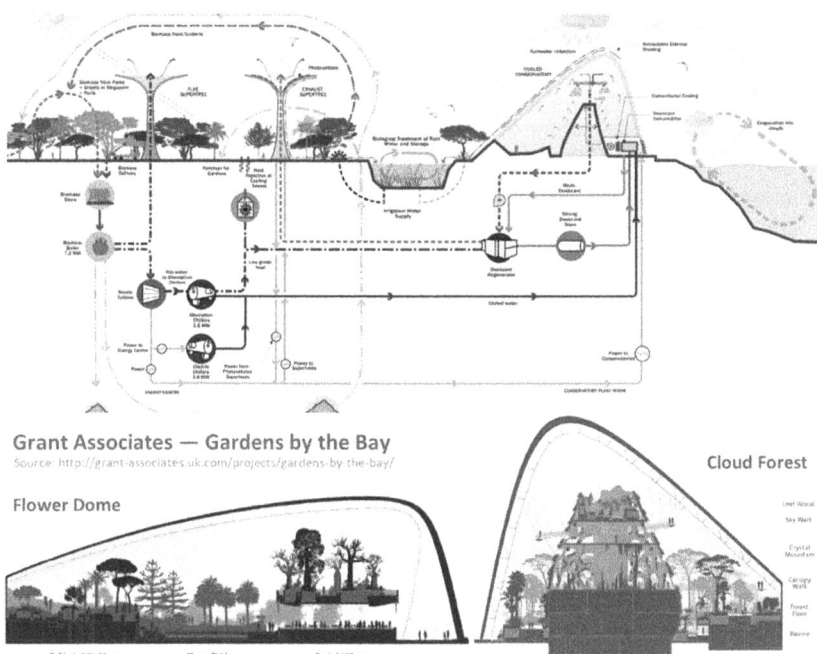

Fig. 4.86: Singapore. Buildings and plant structures of the *Gardens by the Bay* promote sustainable and energy-saving technological systems, allowing to minimise ecological footprint (source: Grant Associates,<http://grant-associates.uk.com/projects/gardens-by-the-bay/>[accessed: 18.11.2018])

Fig. 4.87: Singapore, Kallang River @ Bishan-Ang Mo Kio Park (project by Atelier Dreiseitl, 2012). On the left: the valley with a concrete riverbed of the Kallang River. On the right: the river transformed into a meandering watercourse, source: <https://howlingpixel.com/i-en/Bishan-Ang_Mo_Kio_Park>)

Another section of the river covered by the **Kallang River @ Potong Pasir** project flows in the vicinity of Pong Pasir housing estate and St. Andrew's school complex. Open debate, prior to the investment, enriched the area development concept with rain gardens and bio-retention channels. The solutions aim to retain and pre-purify rainfall runoff, simultaneously raising the students and the inhabitants' awareness of the processes, which increases the youth's engagement in the environment protection (PUB 2018).

Vitality of short Singapore rivers depends on sustainability and purification of surface runoff from highly urbanised catchment areas. The effectiveness of water management results from the scale of innovative implementations. From 2010, when the *ABC Waters* programme was initiated, to 2018 as many as 75 certified projects (PUB 2018) were realised: water gardens and public spaces with green infrastructure[301], as well as riverside parks and areas[302] and innovative facilities of ecological architecture[303].

The rivers of Singapore are regaining vitality owing to natural regeneration of the valleys and urban basins. Natural ecosystems, previously destroyed due to urbanisation, are being reconstructed in buildings (green roofs, terraces and walls, indoor water circulation), in urban public and private space as well as river valleys. The example of Singapore excellently illustrates an efficient way of return, which depends on integration of various fields of knowledge and coordination of actions with a view to shaping productive, vital, beautiful rivers and public spaces.

---

301 E.g. Keat Hong Crest by Housing & Development Board; eCO by Far East Organisation, Frasers Centrepoint Ltd and Sekisui House Singapore Pte Ltd; Skyville @ Dawson by Housing & Development Board; Jurong Eco-Garden by JTC Corporation.

302 E.g. Sengkang Riverside Park by National Parks Board; Waterway Sunray by Housing & Development Board; Waterway Sundew by Housing & Development Board; The Learning Forest at The Singapore Botanic Gardens by National Parks Board (Gold Certified); Rivervale Shores by Housing & Development Board (Gold Certified).

303 E.g. Gardens by the Bay by National Parks Board; Tree House by City Developments Limited/Hong Realty (Pte) Ltd; Khoo Teck Puat Hospital by Alexandra Health Pte Ltd; SkyResidence @ Dawson by Housing & Development Board; Yale-NUS College by National University of Singapore; Singapore Sports Hub by Sports Hub Pte Ltd/ Sport Singapore.

5 Summary

> *In contrast to the Western, utilitarian view of a river as a resource to be exploited by humans, the communities claim distinct relationships with the river based on guardianship, symbiosis and respect, in which the rivers have an intrinsic right to exist. That is the core-concept of bio-cultural rights.*
>
> [Erin O'Donnell 2018]

5.1 Return to friendship

Can relations between cities and rivers return to friendship? Are we ready to notice and respect the rights of rivers, accepting the fact that they constitute living structures to the same degree as our cities?

The book was inspired by the report on the Whanganui River (2017), saint river of the Maori people - native inhabitants of New Zealand, which gained the status of a legal personality on the basis of the New Zealand Parliament act (15th March 2017)[304]. It has been the first but not the only case in modern history when a river was acknowledged as a "living being" and has rights and human representatives protecting its interests. The same year the Ganges and the Jamuna rivers, sacred to Indian inhabitants, were provided with special protection by the Nainital court in the northern state of Uttarakhand, which granted them the status of "living beings" and legal personalities[305]. In 2017 the Constitutional Court of Colombia also granted a status of a legal personality to the Rio Atrato river on the basis of eco-centric approach and the theory of bio-cultural rights (O'Donnell 2018)[306].

---

304 Michalak A., "New Zealand Parliament acknowledged the worshipped by the Maoris river as a legal personality". *Rzeczpospolita*, 16.03.2017 < http://www.rp.pl/Spoleczenstwo/170319159-Nowa-Zelandia-Rzeka-Whanganui-zostala-uznana-za-osobe-prawna.html> [accessed: 10.06.2018].

305 "India: saint rivers gained the status of "living beings" and legal personalities". PAP, 21.03.2017 < https://www.polskieradio.pl/5/3/Artykul/1742259,Indie-swiete-rzeki-otrzymaly-status-bytow-zywych-i-osobowosc-prawna> [accessed: 10.06.2018].

306 Erin O'Donnell in the book *Legal Rights for Rivers* analyses functions of *environmental water managers* (EWM) as well as possibilities opening for representatives resulting

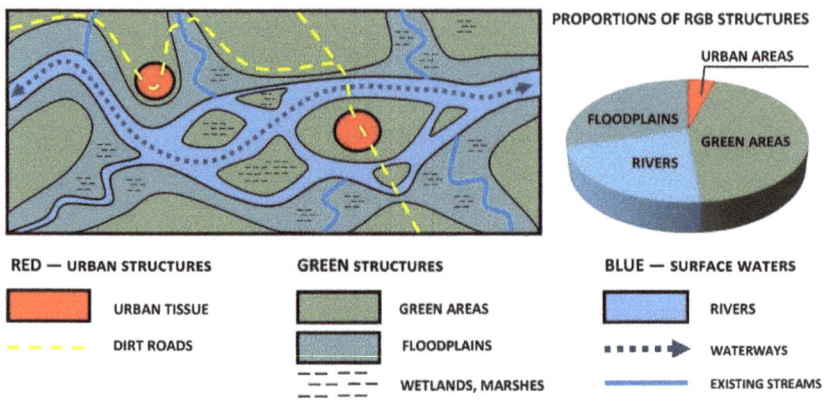

Fig. 5.88: A diagram of spatial relations between a city and the environment as well as ratio of RGB structures (pie chart) in the RESPECT period (developed by A. Januchta-Szostak)

The rights of nature (Nash 1989, Boyd 2017) and empowerment of rivers brings technocrats' ironic smile. Techno-economical governance of water seems to be much less complicated. Treating rivers as living ecosystems evokes moral discomfort with the scale of misuse during their conquest and exploitation. Drawing conclusions from the periods of RESPECT, CONQUEST and RETURN may help us "open our eyes" and notice the interdependencies between systems of social values, priorities of cities' development and their consequences in river valleys, between urbanisation and escalation of threats, between return to respect and intergenerational responsibility.

The **RESPECT** period, which lasted from the beginnings of settlement to great geographical "discoveries" at the turn of the 15th century, featured balance between use of rivers for human needs and regenerative capabilities of the natural environment. Admittedly, the level of technical knowledge made it impossible to control the element but also the scale of spatial interference was so insignificant that it did not cause environment degradation (Fig. 5.88). The nature could survive even such great undertakings of the ancient era as irrigational systems of Egypt and Mesopotamia, Chinese navigational canals, urban water supply and sewage systems of Mohenjo-Daro as well as sewage discharge in medieval cities. Riverside human settlements gained defensive, transport, economic and

---

from the rivers' status of legal personalities (e.g. a possibility of concluding contracts or defending rights in court).

cultural advantages from water but they did not enter the flood plains. Whereas the resilience and dynamics of natural systems effectively compensated for the consequences of anthropogenic transformations (Wojciechowski 2000).

According to stoicism, the Nature was intelligent, harmonious and divine while the sanctity of water, embodied in the pantheon of deities, was worshipped[307]. Simultaneously, ancient scientists looked for the principles of hydrodynamics while constructors implemented hydro-technical inventions (channels, aqueducts, bridges, harbours etc.). Maybe then it was not technological helplessness but rather original wisdom which made people respect the element? A debate on changing the course of some tributaries of the Tiber held in the Roman Senate in 15 AD is an example of a cautious approach as the delegates pointed out: "it is the nature itself which has taken the best care also of people's welfare and fate, determining the springs, riverbeds and estuaries of rivers" (Krawczuk 1998, p. 102). Admittedly, in medieval times Christian religion deprived rivers of sanctity, though technological regress did not allow to cease to respect the power of water. The Dutch created a unique culture of coexistence with water, combining respect with expansion. Draining another polder, they were aware of the price they might pay for every interference and of the fact that controlling the element requires constant efforts of human community.

The period of **CONQUEST**, dating from the 16th century to the 1970s, actually began much earlier and is still ongoing in many countries.

The path for conquest was gradually cleared by hydro-technical inventions and imperial expansion of Rome. Whereas in the Middle Ages, anthropocentrism based on religious grounds began to prevail in the Europeans' awareness, which determined the direction for the conquest of nature and non-Christian cultures in the period of the great geographical "discoveries". Actions orientated towards subjugation and exploitation of raw materials and human resources inseparably accompanied colonial policy pursued by European countries until mid-20th century[308]. The conquest period (Fig. 5.89) featured subjugation approach to the

---

307 Nota bene: the pantheon of Greek and Roman gods personifying the forces of the Nature, remained in all fields of European modern culture, despite depriving the nature of sanctity status.
308 According to Marek Kosmala (2011, p. 5) the period of subjugation and exploitation lasted from the beginning of the 19th century to the second half of the 20th century; but, in my opinion, crucial changes in relations between humans and the Nature already occurred at the beginning of colonialism.

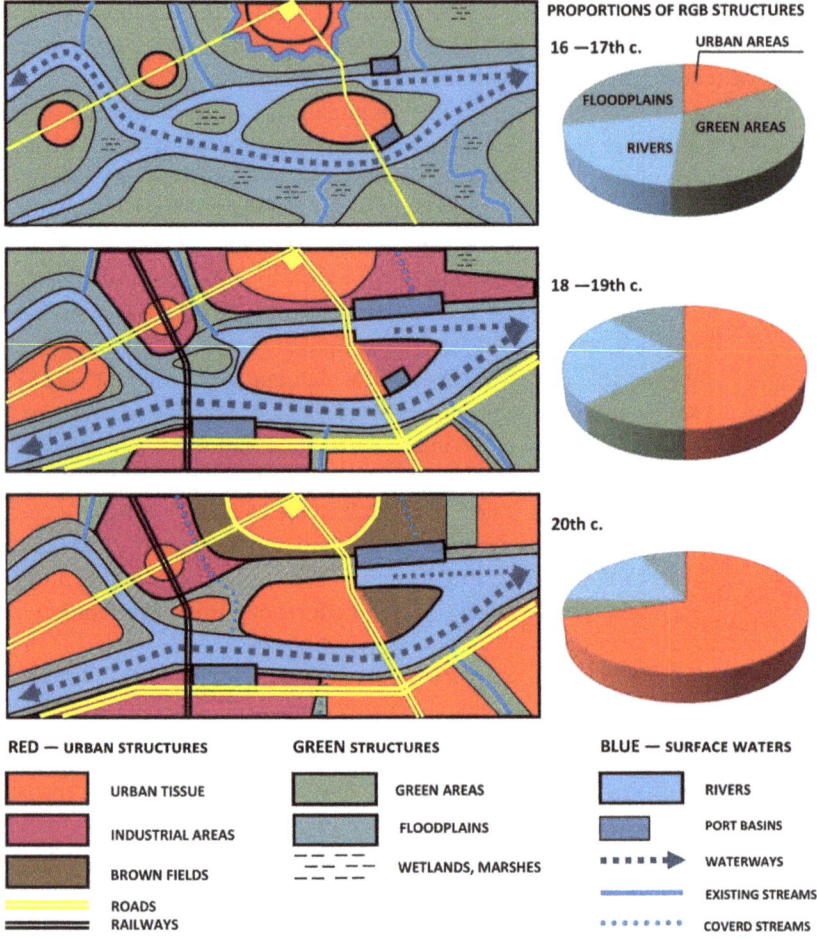

Fig. 5.89: A diagram of spatial relations between a city and the environment as well as ratio of RGB structures (pie chart) in the CONQUEST period (developed by A. Januchta-Szostak)

environment and to the society, originally justified by "the divine right of power" and subsequently by utilitarianism and economic liberalism.

Contra-posing culture and nature led to considering the level of subjugation and "beautification" of the nature as the indicator of progress. Regulation and exploitation of rivers enabled the development of inland navigation and trade, which was the foundation for the power of harbour cities in colonial empires.

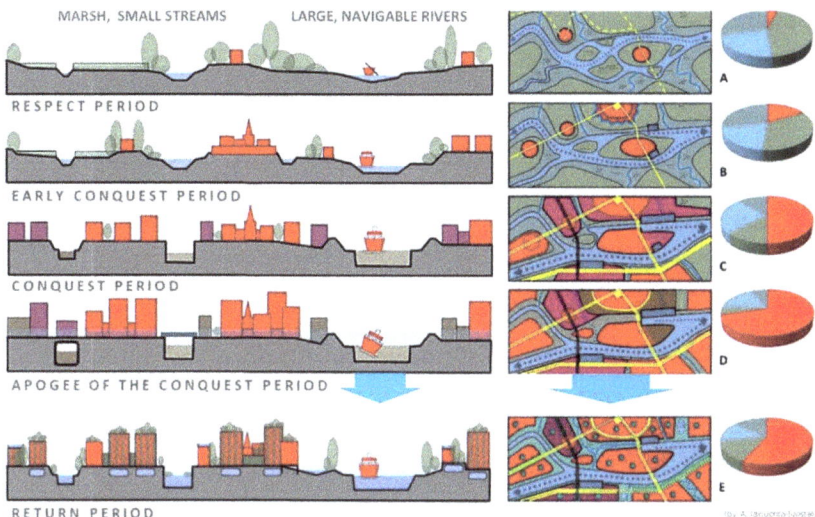

Fig. 5.90: A comparative profile: diagrams of changes in river valleys intersection and ratio of RGB structures in the urbanisation process (developed by A. Januchta-Szostak)

While the achievements of science and technology at the end of the 18th century eventuated in the Industrial Revolution. Despite the undeniable economic benefits, resulting from the technological progress, the euphoria made us forget about the limitations of "organic humanism", pointed out by Lewis Mumford in his book titled *The Condition of Man* alredy in 1944.

The 19th century industrialisation of riverside areas and chaotic city expansion not only led to degradation of rivers whose role was reduced to transport routes, sources of water supply and sewage collectors. The scale of urbanisation and resources exploitation in 20th century eventuated in disastrous aftermath for the inhabitants of the whole Earth. Regrettably, the effects of these actions appear with a significant delay and affect future generations. Only looking from the perspective of hundreds of years allows to notice the dramatic changes in the ratio of built (R), natural (G) and water (B) structures in cities (Fig. 5.90).

In the conquest period, big European rivers (e.g. the Rhine, the Danube and the Meuse) were regulated for the purposes of navigation, hydro-power and flood protection; the floodplains were separated by embankments and developed, the swamplands – drained, small urban watercourses – transformed into sewage collectors, vast areas of urban ground – sealed and drained while the rainfall runoff – directly discharged into closed, underground sewage systems.

Summary

How much space have we left for water in urban catchment areas? Why are we still surprised by urban floods? What are the self-regeneration chances of urban rivers separated from their ecosystem sewage treatment plants? And finally: what living conditions are offered by concrete structures devoid of greenery and living rivers? The questions and the noticeable consequences of the environment degradation, both at a global and city scale, have facilitated the awareness raising process which has changed the direction for civilisation evolution and guided us to the way of RETURN.

The period of **RETURN** – regaining the lost relations with rivers and restoring respect for the environment and natural resources, which began in the 1970s, has been undergoing subsequent shifts in developmental priorities (Fig. 5.91).

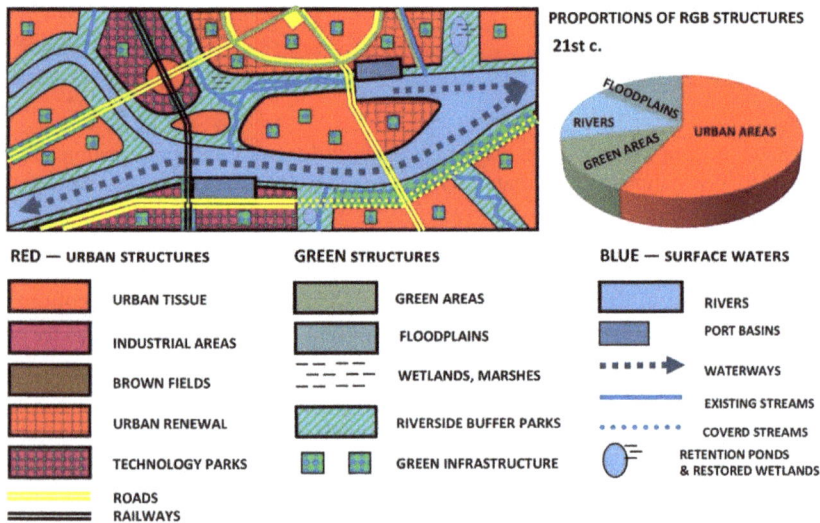

Fig. 5.91: A diagram of spatial relations between a city and the environment as well as the ratio of RGB structures (pie chart) in the RETURN period (developed by A. Januchta-Szostak)

Return to large rivers, crucial for cities identity, has become the main slogan of urban waterfront renewal strategies in many cities (e.g. Rotterdam, Hamburg or Warsaw). Socio-economic needs, like reclaiming the degraded post-industrial terrains for business and housing as well as creating conditions for leisure in river valleys, became the catalyst for waterfront transformation. However, high level of rivers pollution and their urban tributaries along with lack of green structures continuity required performing wider regenerative actions. The period of "green breakthrough" emphasised ecological aims[309] which drew town planners' attention to reconstructing and boosting eco-hydrographical systems, uncovering small watercourses and their re-naturalisation (e.g. in London).

The paradigm shift in water management and governance (Directive 2000/60/EC) and flood risk management (Directive 2007/60/EC) and their integration with urban planning and governance occurred relatively late: in the 21st century. Considering global climate change, the transformation appeared urgent. The increase in flood threats prompted the EU states to implement programmes on making space for water in whole regions.

Nevertheless, planning principles were only modified in cities when the relationship between the transformation of urban catchments and floods in the valleys could no longer be ignored. Currently, European cities implement strategies, programmes and systems allowing mitigation and adaptation to climate change. Integrated urban development strategies (ISUDS) including integrated urban water management (IUWM) and catchment-based approach (CaBA) enable reconstruction of retention potential and environmental vitality with the us of blue-green infrastructure. However, new values are still breaking through the resistance of technocratic habits and current interests as well as social conformism and scepticism.

Resistance to actions that entail high environmental cost is seen as "eco-terrorism" by some public circles, which is accurately illustrated by a comment found online: "I'm not for destroying the nature, at the end of the day everybody appreciates its scenery for walks etc., but treating the nature as the Absolute, a deity whose rights are not supposed to be questioned is another idea of Neo-Luddites, anarcho-primitivism followers and Unabomber's apologists along with his sick primitive obscurantist troglodit attitude"[310].

In the 19th century, Sitting Bull (1831–1890), the chief of Hunkpapa Lakota tribe, apparently said "When the last tree has been cut down, the last river poisoned and the last fish caught – only then will we realise that one cannot eat money". In the 20th century we discovered global scale of environment destruction, alarmed by the reports by Sithu U Thant (1969), Brundtland Commission

---

309 I.a. significance of wetlands (Ramsar Convention 1971), the necessity of the environment protection (Stockholm Declaration, 1972) and sustainable development (Rio Declaration and Agenda 21, 1996).

310 Translation of the comment by KapitanKorsarzPirat 27th February 2018 at 9:30 <https://demotywatory.pl/4835354/Kiedy-wyciete-zostanie-ostatnie-drzewo-ostatnia-rzeka-zostanie>

(1987) and IPCC; nevertheless, the necessity of limitations of fossil fuel and greenhouse gases is contested in some political circles. "Despite definite increase in the convincing scientific proof (...), a phenomenon known as climatic scepticism still prevails" (Kundzewicz 2018, p. 160), resulting in short-sighted energy policy based on exploitation of limited coal resources. Even being fully aware of circular interdependencies in the cycle: energy – climate – water governance – urban planning – energy, we still choose interim goals and instant effects in our local "yard", following political terms or billing cycles. The trends initiated in Europe and in the world during the RETURN period are barely followed by Polish spatial and water policy standards, though a number of local initiatives[311] contribute to popularisation of *nature based solutions* – NBS.

5.2 Vital rivers

Regulated rivers, canalised or hidden underground, devoid of natural hydromorphological processes and biological life, which I described in the second chapter, are still common in our cities. Some, like the Rawa river in Chorzów and Świętochłowice, were buried just a few years ago (in 2010). "After works transforming the Rawa into "a pipeline", the contractor flattened the area of the former river course and planted it with greenery. [...] Thus the Rawa will never be an open river – since it is not possible to discharge the sewage in any other way than into its course. The inhabitants gained walking areas in the place of the former Rawa and soon will not remember they are strolling on the forgotten river".[312] The programmes implemented in London (cf. Chap. 4.3) and Singapore (cf. Chap. 4.4) as well as numerous examples of watercourses regeneration, described in the third chapter (cf. Chap. 3.3) prove we do not forget so easily. The analysis of the benefits and costs of covering the Rawa may not have considered the lost ecosystem benefits as well as the expenses of the watercourse restoration which will be incurred by the future generation.

The services of river ecosystems, i.e. benefits from the nature (Kronenberg & Bergier 2010), include not only indispensable functions of supply, habitat and

---

311 E.g. restoration of greenery and creating blue-green infrastructure in Wrocław include: "Zieleń bez granic" project (Greenery without borders), "GROW GREEN" project connected with adaptation to climate changes through building pocket parks, "Szare na zielone" (Grey into green) orientated towards ecological education for schools, and many others.

312 Source: <http://swietochlowicki.pl/aktualnosci,h,23,1,r,1079.html> [accessed: 20.07.2018].

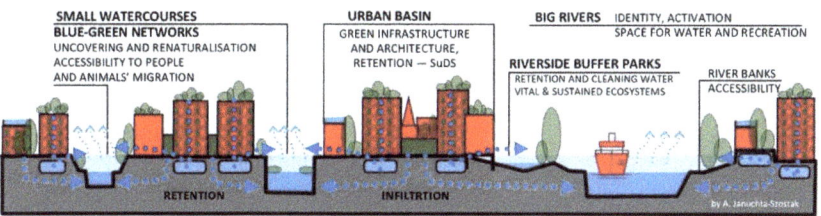

Fig. 5.92: Revitalising rivers requires actions in valleys, tributaries systems and the entire catchment area, in all RGB structures (developed by A. Januchta-Szostak)

regulation but also a number of cultural services concerning leisure, tourism, recharging batteries and experiencing beauty, which determine life quality in a city. Reconstruction of rivers vitality (Fig. 5.92) requires then multidimensional actions in valleys, tributaries systems and the entire catchment areas (Tab. 5.4), orientated towards:

- **cultural revitalisation (R)**, connected with regaining identity of rivers, invigorating riverside public spaces, enhancement of their accessibility and landscape quality, but also ecological education and developing social attitudes of co-responsibility for water;
- **natural regeneration (G)**, making it possible to reconstruct continuity and vitality of eco-hydrographical structures and ecosystems productivity in river valleys and whole catchment areas, whose development influences the quality and quantity of rainfall runoff supplying the rivers;
- **hydrological sustainability (B)**, which determines watercourses flow dynamics, water erosion intensity and flood hazard level, requires then increasing the space for water in valleys of big rivers, in tributaries network as well as raising retention capabilities of urban catchment areas.

Architects and town planners most often focus on urban waterfront areas whose transformation brings the most spectacular image effects. It challenges solutions concerning raising the landscape aesthetics (often hydro-technical development and anti-flood protection systems), negotiating borders as well as visual integration and functional coexistence between water and architecture (Nyka 2013). The opportunity for creating attractive public spaces in linear riverside structures or compositional axes perpendicular to the river enable search for finding connections between the city and the river (Kosiński 2011). Unlocking the potential of multi-functional, flexible solutions, makes it possible to "recycle" urban space and adapt the waterfront to new needs, including increasingly more

Tab. 5.4: The template of cultural (R), environmental (G) and hydrological (B) goals orientated towards invigorating urban rivers (developed by A. Januchta-Szostak)

	LARGE RIVER	URBAN TRIBUTARIES	URBAN CATCHMENT
R	Identity reconstruction, Visual and functional connection with the city; Buildings facing the river – landscape quality; Accessibility of the banks; Activation of river and waterfront public space;	Uncovering and re-naturalisation of canalised watercourses – landscape quality; Accessibility and continuity of water network – creating areas and leisure routes along watercourses; Using landscape advantages of water;	Ecological education – awareness of water circulation, respect for the resources and rights of water; Developing co-responsibility for water; Circular and integrated urban water management (IUWM);
G	Clearing valley migration corridors; Vitalising waterside ecosystems, especially littoral zones, riverside buffer parks;	Reconstruction of eco-hydrological structures – blue-green networks (quality and continuity); Enhancement of biodiversity and vitality of ecosystems;	Increasing biologically active areas and greenery in cities; Green architecture & infrastructure;
B	Space for water - restoring valleys retention capacity; Reclaiming flood areas; Limiting barriers (dams, barrages, weirs); Enhancement of water condition;	Recovering natural hydro-morphological processes; Restoring wetlands; Flexible flood protection systems; Good ecological status or potential of waters;	Catchment-based approach in urbanism (WSUD) – reducing the surface runoff rate, water retention; Sustainable systems of rainwater management SuDS; Reclaiming hydrological cycle;

important leisure and ecosystem functions. The example of Singapore, provided in the fourth chapter, proves that "green" town planning, architecture and infrastructure integrated with water management give even more opportunities of creating space which is friendly to people, environment and water.

Even not referring to the whole spectre of human rights, which the status of "legal personality" grants, it is difficult to refuse rivers **the right to live** while maintaining natural biological and hydro-morphological processes, which have been noticed and guaranteed in Europe by environmental and water directives. Also urban rivers deserve the right to be clean and self-purifying through connection with ecosystems, the right to flow and be included in natural hydrological cycle or the right to overflow or flood in accordance with the hydrological laws of nature like all the other watercourses. The significance of rivers is also deeply rooted in culture. Their character, beauty and influence on diverse areas of individuals and communities' lives, including development of cities, require acknowledging their identity and landscape distinctiveness. In order to "evaluate the intangible": beauty, silence, emotions and spiritual experiences, which are elements of non-material services of water ecosystems, it is necessary that the principles of the economy of values should be applied and not the measurement of monetary value (Hausner 2017).

5.3 Urbanism and water

The history of relations between cities and water can be also perceived from the perspective of the evolution of needs and challenges which cities had to face at the subsequent stages of civilisation development (Brown et al 2009, Fig. 5.93). Those factors determined the range and methods of water sources use as well as linkage between town planning and water governance.

Water supply (Fig. 5.93, phase 1) and **sewerage** have been the first and fundamental needs of cities. Admittedly, water supply-sewage systems existed already 4.5 thousand years ago (e.g. Mohenjo-Daro approx. 2550 BC) but the majority of European cities were sewered as late as in the second half of the 19th century, having experienced mass epidemics caused by water contamination (Fig. 5.93, phase 2). Contemporary mega-cities of over 10 m residents in developing countries still struggle to meet the inhabitants' basic needs.

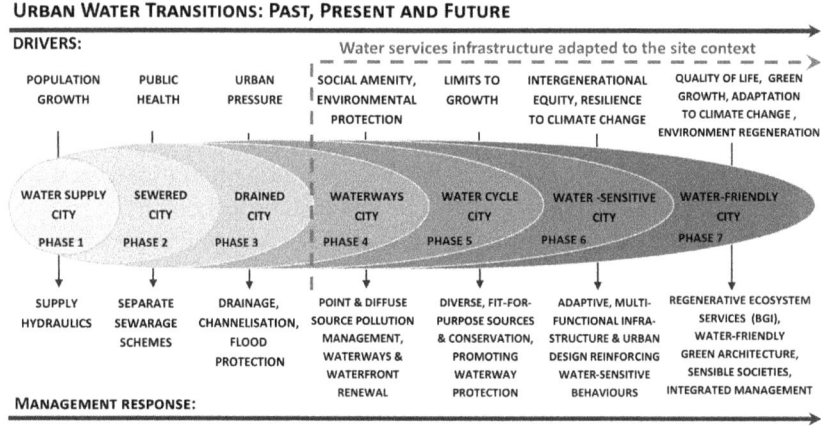

Fig. 5.93: Phases of changes in relations between a city and water, determined by key development factors (developed by A. Januchta-Szostak based on Brown et al 2009)

Housing and infrastructure development was performed at the cost of biologically active areas and water structures, so cities in developed countries have been excessively sealed and drained. Swamplands, backwaters, former moats, canals and even small watercourses and reservoirs were eliminated which led to extreme dehydration of cities (Fig. 5.93, phase 3); while the water issue disappeared from the scope of town planners' competence and responsibility (Shannon and De Meulder 2008).

Social and landscape water appeal has been the catalyst for revitalisation processes of riverside areas, initiated in Europe in the 1980s. **Water urbanism** (Shannon et al 2008; Nyka 2013) covers the areas where the urbanised structures and water meet, so **cities on waterways** (Fig. 5.93, phase 4), driven by the development of navigation, have turned their facades towards their rivers. The effectiveness of the return to rivers depended though on the environment quality and the surface water condition which was emphasised by "eco-urbanism" (*Eco-urbanism* – Ruano 1998) and "green urbanism" (*Green urbanism* – Beatley 2000).

The expansion of cities in the 20th and 21st centuries faced barriers of water supply problems and sewage treatment, thus city authorities have to consider **cirlular water management** (Fig. 5.93, phase 5) in the subsequent development phase. Whereas town planning had to manage to not only restore visual and functional relations with the river but also analyse eco-hydrological consequences of urbanisation as well as minimisation of negative impact of built environment and sealing on the natural environment (Feyen et al 2009; Januchta-Szostak 2012).

The awareness of problems and water determinants in urban planning and design increased already at the turn of the 20th and 21st century which led to including **the water-sensitive urban design** (WSUD, France 2002) in the canon of modern design trends. The need for integration of water management with spatial planning and environment protection is interrelated with climate challenges and intergenerational responsibility. In **water-sensitive cities** (Fig. 5.93, phase 6) public space and built environment are shaped using multi-functional spatial and technological solutions (SuDS, LID, WSUD), applying both grey and blue-green infrastructure; gradually, also eco-hydrographical structures in urban catchments are being restored.

Intensification of urbanisation processes, global climate change and perspective of water crisis pose new challenges to town planning. A subsequent but surely not the last stage might be **"water-friendly city"** (Fig. 5.93, phase 7), which will satisfy the requirements of high quality of life and green growth but will also face climate changes and meet the needs for regeneration and increasing environment productivity, energy effectiveness and circular economy. In order to achieve the phase, the principles of "sustainable development" need to be changed into "regenerative development" (Lyle 1994, Pedersen Zari 2012), according to which environmental life support systems constitute the basis and the goal of shaping cities with the use of NBS and the principles of ecosystems dynamics. The goals of **water-responsible urbanism** are orientated not so much towards mitigation of negative consequences of anthropogenic transformations but towards "positive development" (Birkeland 2008), achieving synergy effect in symbiosis between the city and the nature and their one joint hydrographical "bloodstream" (Fig. 4.55). Reaching the stage of friendship with water requires changing the system of social values, developing the attitude of responsibility for vitality of urban ecosystems and integrated management which respects rivers identity. The vision of symbiotic, biological-cultural urban structures seems hardly realistic but it has already blossomed in town planners' imagination and its implementation is only a question of our determination.

Abstract

River-friendly cities

The history of urbanisation was inseparably connected with the use of the environment, the transformation of the catchment areas as well as the subjugation and exploitation of rivers. Today we experience the effects of this expansion in the form of escalating water problems, which are serious threats to humanity in the era of global climate change and the continuous growth of the urban population. The book outlines the processes of transformation of urban (Red), natural (Green) and waterborne (Blue) structures in three phases of civilisation development: the period of RESPECT, CONQUEST and RETURN.

The period of RESPECT (from the beginning of the settlement to the 16th century) was characterised by sustainable use of resources, allowing the natural environment to regenerate naturally. The cities were balancing on the edge of benefits and threats, on the one hand using transport, defence and production assets of water, on the other hand, respecting the flood risks. A distinctive feature of this period was also subjective or even sacral approach to rivers, resulting not only from the peoples' dependence on water and the lack of technical capabilities of resistance, but also from the holistic perception of nature.

The CONQUEST period (from the 16th century to the 1970s) began in the era of great geographical "discoveries" which introduced Europe into the era of colonialism, entailing exploitation of natural resources and non-Christian native cultures. The economic growth, scientific and technical progress became a catalyst for the industrial revolution and urban expansion as well as further subjugation of rivers. The scale of anthropopression in the 19th and 20th centuries has changed the Earth's environment and climate. Therefore, in the Anthropocene epoch, cities have to face the intensification of hydro-meteorological extremes.

The period of RETURN to the rivers began in Europe in the 1970s with the revitalisation of degraded urban waterfronts (e.g. Kop van Zuid in Rotterdam, Docklands in London). Raising ecological awareness and social pressure during the "green turning" of the 1970–80s led to gradual improvement in surface water quality. Further goals included flood risk management, river restoration, regeneration of water resources and the environment of urban catchment areas by creating blue-green infrastructure and sustainable rainwater management systems. Integrated urban water management (IUWM) includes also institutional cooperation aimed at improving the cities' resilience to the hydrological consequences of climate change. The strategies of RETURN to river-friendly relations have been illustrated by examples of Rotterdam, London and Singapore – cities with a vision based on an integrated management of space, environment and water, applied with the help of effective strategy planning and implementation tools.

The water-responsible urbanism in one of the challenges of 21st century. The return to friendly relations between cities and rivers demands holistic management of urban space and the restoration of respect for the environment. The effectiveness of the RETURN depends on the integration of various fields of knowledge and coordination of activities in order to shape sustainable and multi-functional urban spaces combining red-green-blue goals and structures.

List of figures

Fig. 1.1: Prague, 1608, Abraham Saur, *https://www.vintage-maps. com/>* [accessed: 18.11.2018] 26

Fig. 1.2: Vienna, 1608, Abraham Saur, *https://www.vintage-maps. com/>* [accessed: 18.11.2018] 27

Fig. 1.3: A plan of Koln, 1800, John Andrews, *http://www. ancestryimages.com/>* [accessed: 18.11.2018] 27

Fig. 1.4: A plan of Paris, 1776/1800, John Andrews, *http://www. ancestryimages.com/>* [accessed: 18.11.2018] 26

Fig. 1.5: Lübeck, 1910. Own study based on: *http://www.lib.utexas. edu/maps/map_sites/hist_sites.html>* - Courtesy of the University of Texas Libraries, The University of Texas at Austin ... 35

Fig. 1.6: Gdańsk, 1883. Own study based on: UMGEBUNG VON DANZIG, Baedeker Karl, Mittel- und Nord-Deutschland, ed. GeographischeAnstalt von Wagner & Debes, Leipzig 1883 36

Fig. 2.7: The plan of an ideal harbour city by Simon Stevin (1590), source: Kostof 1991 50

Fig. 2.8: Dutch topographical map of a polder Haarlemmermeer (2015), source: Janwillemvanaalst, [access: 13.01.2019] 51

Fig. 2.9: The transformation stages of Zuiderzee Bay. Source: Hooimeijer et al. 2005 51

Fig. 2.10: The stages of Amsterdam development. Developed on the basis of historical maps and the book: Hooimeijer et al 2005 52

Fig. 2.11: A map illustrating the percentage of heavily modified and artificial water bodies in river basin districts. Source: Map of percentage of heavily modified water bodies and artificial water bodies in River Basin Districts Version 29 October 2012 *http://ec.europa.eu/environment/water/ water-framework/facts_figures/pdf/Heavily%20modified%20 water%20bodies-2012.pdf>* [accessed: 30.07.2018] 60

Fig. 2.12: The Danube valley in Vienna in 1867, before the main engineering works. The dark line – the main course of the Danube in accordance with the planned regulation. Source: (Donauhochwasserschutz 2017) *https://www.wien. gv.at/umwelt/gewaesser/pdf/donau-hochwasserschutz-2017. pdf>* [accessed: 4.08.3018] 64

Fig. 2.13: The Danube and its tributaries (1–8) in Vienna after the regulation in 1875. (layout by: Friedrich Hauer, based on the hydrographic map of Magistratsabteilung 45 "Wiener Gewässer", city of Vienna)*https://link.springer.com/article/10.1007/s12685-013-0079-x*> [accessed: 3.08.2018] 66

Fig. 2.14: The engineering of the Danube in Vienna. Source: *Danube River in Vienna 1529–2010* (Wiener Donau 1529–2010)*https://www.youtube.com/watch?v=dHERpWgA84Y*> [accessed: 3.08.2018] based on the ENVIEDAN research project Nr: P 22265-G18, under the supervision of: Univ. Prof. Ing. Dr.phil. Verena Winiwarter. *http://www.umweltgeschichte.uni-klu.ac.at/index,3560,ENVIEDAN.html* 67

Fig. 2.15: A comparative profile of Poznań hydrographical structure at the end of the 17th century and a contemporary picture. (developed by A. Januchta-Szostak based on historical map developed by P. Biskupski – on the left and contemporary Google map – on the right) 72

Fig. 2.16: The hydrographical system of Łódź: the rivers entirely or partially canalised (coloured dark blue) and non-existent sections of rivers (coloured light blue). Based on the study by M. Stolińska (2018) and source maps Source:*http://mapa.lodz.pl/portal/apps/webappviewer/index.html?id=7489987eb60a4369814e53d49cc58ffc&extent=19.317,51.7126,19.6459,51.8507*>[accessed: 1.06.2018] 73

Fig. 2.17: The Pełtew River buried in the underground of Lwów. Photo: Vladislav Vozniuck, source: <https://general-kosmosa.livejournal.com/61009.html> [accessed: 1.08.2018]. 74

Fig. 2.18: The view of the development along the La Bièvre River, the tributary of the Seine in Paris (la rue des Gobelins) before the redevelopment by Haussmann in the second half of the 19th century, source: Crossman, 2013. 74

Fig. 2.19: A comparative profile of Paris plan from the mid-16th century (on the left) and part of a plan from the turn of the 18th century (on the right) 80

Fig. 2.20: Part of Paris plan from 1875, after the redevelopment by E. Haussman 81

Fig. 3.21: The number of bi- and multilateral environment agreements in Europe since the beg. of the 20th century, developed by A. Januchta-Szostak based on: EEA, 2010, *Environmental Agreements since 1900*, source: https://www.eea.europa.eu/

	data-and-maps/figures/environmental-agreementssince-1900> [accessed: 5.08.2018].	97
Fig. 3.22:	Astrup Fearney Museet at Aker Brygge in Oslo (photo by A. Januchta-Szostak)	104
Fig. 3.23:	Former harbour Aker Brygge waterfront in Oslo (photo by A. Januchta-Szostak)	104
Fig. 3.24:	New development of Zeeburg district in Amsterdam - Jawa (photo by A. Januchta-Szostak)	104
Fig. 3.25:	"NEMO", New Metropolis in the Eastern Dock refers to harbour past of Amsterdam (photo by A. Januchta-Szostak)	105
Fig. 3.26:	Park of the Nations by the Tagus in Lisbon with hardly accessible bank line (photo by A. Januchta-Szostak)	105
Fig. 3.27:	The European Solidarity Centre (project by FORT) - the icon of the Young City in Gdańsk (photo by A. Januchta-Szostak)	107
Fig. 3.28:	The revitalisation of the Motława River waterfront in Gdańsk - Wyspa Spichrzów (photo by A. Januchta-Szostak)	109
Fig. 3.29:	Xawery Dunikowski Waterfront in Wrocław (photo by A. Januchta-Szostak)	110
Fig. 3.30:	Mill Island in Bydgoszcz and reconstructed canal of Międzywodzie (photo by A. Januchta-Szostak)	111
Fig. 3.31:	The semi-natural, right Vistula riverbank in Warsaw – a valuable natural area of Natura 2000 in the centre of the capital – the world phenomenon and a unique space of contact with the nature for the inhabitants of the big agglomeration (photo by A. Januchta-Szostak)	112
Fig. 3.32:	Warsaw: the left (on the left) and the right (on the right) Vistula bank viewed from Poniatowski bridge (photo by A. Januchta-Szostak)	113
Fig. 3.33:	Warsaw – boulevards by the Vistula. On the left: Kościuszkowski waterfront – a narrow bank zone separated from the city by a communication artery. On the right: the new section of the Vistula Boulevards – Powiśle (photo by A. Januchta-Szostak)	113
Fig. 3.34:	Cracow: the Vistula meander by Wawel – comparative profile by courtesy of R. Konieczny. On the left: a historical photo: Digital National Museum (1-G-4566–2). On the right: modern view (photo by R. Konieczny)	114
Fig. 3.35:	Percentage of water bodies not in good ecological status or potential per river basin district (RBD). Source: (EEA 2018) https://www.eea.europa.eu/highlights/european-waters-getting-cleaner-but> [accessed: 7.08.2018]	117

Fig. 3.36: The number and types of entities involved in the projects on regeneration of rivers in Europe from 1989 to 2016 (Szałkiewicz 2018) 117

Fig. 3.37: The Isar river before the redevelopment (on the left) and after the re-naturalisation (on the right). Photo by Daniela Schaufuß; source: Munich city, *https://climate-adapt.eea. europa.eu/metadata/case-studies/isar-plan-2013-water-management-plan-and-restoration-of-the-isar-river-munich-germany*> [accessed: 7.09.2018] 121

Fig. 3.38: Anam-ro/Seognbukcheon, Seoul, South Korea (URB-I n.d.). Above: the view before uncovering the watercourse. Below: after the restoration 125

Fig. 3.39: Suwon cheon, Suwon, South Korea (URB-I n.d.).Above: the view before uncovering the watercourse. Below: after the restoration 125

Fig. 3.40: Shinan-gil, Daejon, South Korea (URB-I n.d.). Above: before, below: after the changes 126

Fig. 3.41: Noordwal/Veenkade, the Hague, Holland (URB-I n.d.). Above: before, below: after the changes 126

Fig. 3.42: Sint-Jacobsnieuwstraat, Ghent, Belgium(URB-I n.d.). Above: before, below: after the changes 127

Fig. 3.43: Quai du Jeu de Paume, Chambéry, France(URB-I n.d.). Above: before, below: after the changes 127

Fig. 3.44: Seul, uncovering the Cheonggyecheon stream. On the left: the view before, on the right: after the stream restoration (LAF n.d., Cheonggyecheon Stream Restoration Project) 129

Fig. 3.45: Shanghai, Houtan Park by the Huangpu river. On the left: the view before, on the right: after the waterfront regeneration (LAF n.d., Houtan Park) 130

Fig. 3.46: Leipzig, regeneration of the Pleißemühlgraben canal. On the left: the views before revitalisation. On the right: uncovered and regenerated sections (photo by A. Januchta-Szostak, photos from the 1990s before revitalisation; REURIS 2011) 131

Fig. 3.47: Madrid, the Manzanares river – natural regeneration of the riverbed (photo by P. Nawrocki) 132

Fig. 3.48: Blue-green structure of London requires restoration of valley eco-corridors. Source: ECO CORRIDORS IN LONDON, *http://ecocorridorsinlondon.blogspot.com/*> [accessed: 12.11.2018] 134

Fig. 3.49: Location of the Sokołówka valley with components of Blue-Green Network in Łódź in the background (in accordance with Zalewski 2010). The map developed by: M. Stolarska, source: Krauze et al 2010, p. 35 134

Fig. 3.50: Flood protection strategies: A. Keeping water away from people (e.g. embankments), B. Moving people away from water (e.g. ban on floodplains development), C. Coexistence with water (e.g. non-defensive urbanisation methods), D. Preventive actions in the entire catchment area. Developed by A. Januchta-Szostak 137

Fig. 3.51: Quantitative and qualitative flood threats in the current catchment area system (developed by A. Januchta-Szostak, partially based on Nachlik 2006) 141

Fig. 3.52: Integrated approach to quantitative and qualitative flood threats in catchment area system (developed by A. Januchta-Szostak) 142

Fig. 3.53: Green infrastructure for retention and reuse of rainwater – a concept for Rataje district in Poznań developed under the supervision of A. Januchta-Szostak (the authors of the study: Izabela Jęczmyk, Məşədi Məcid Cavadzadə, Marta Cicha, WAPP 2016) 151

Fig. 3.54: Integrated urban water management IUWM (developed by A. Januchta-Szostak, based on: Bahri 2015 p. 69) 155

Fig. 4.55: Symbolic RGB structure model of sustainable city based on bionics - *Nature-Based Solutions* (developed by A. Januchta-Szostak) 160

Fig. 4.56: Development of Rotterdam at the beg. of the settlement 1000–1340 (Hooimeijer et al 2005) and industrialisation in the 19th century (developed by A. Januchta-Szostak on the basis of historical plans: A.H. Krap 1837 and W.N. Rose 1854 161

Fig. 4.57: Comparison of urbanisation scale of Rotterdam along with the adjoining towns and harbour structures from the mid-19th century to the 21st century (developed by A. Januchta-Szostak on the basis of: Hooimeijer et al 2005) 162

Fig. 4.58: The current view of Rotterdam and its harbour structures. The current view of Rotterdam and its harbour structures. The dashed line marks the modern borders of the city of Rotterdam in the context of the extensive port structures (see figure below) (developed by A. Januchta-Szostak based on of Rotterdam Waterstad. . . 2005) 162

Fig. 4.59: The concept of Kop van Zuid development (project: TeunKoolhaas, 1996). Developed by A. Januchta-Szostak based on: Meyer 1999, p. 355 164

Fig. 4.60: The waterfronts of the Meuse in Rotterdam with the development of Katendrecht, Wilhelminapier and Kop van Zuid in 2007 (photo A. Januchta-Szostak) 165

Fig. 4.61: Rotterdam: HAL headquarters (Holland America Line) on Wilhelminapier. On the left: photo from 1959 (by C. Oorthuys, source: Meyer 1999). In the middle: 2007 (photo by A. Januchta-Szostak) – historical HAL building surrounded by new landmarks: Erasmus Bridge, World Port Centre and Montevideo. On the right: photo from 2018 (by 85martijnh) 167

Fig. 4.62: Transatlantic and the Erasmus Bridge (project B. Van Berkel) – the old and new symbol of Rotterdam. On the left: photo by A. Januchta-Szostak (2007). On the right: photo by KERSTIN B (2018) 168

Fig. 4.63: Rotterdam is threatened with sea, river, ground and precipitation waters. The figure illustrates flood threads and the elements of technical infrastructure used to protect and prevent floods (developed by A. Januchta-Szostak) 224 169

Fig. 4.64: The catchment-based approach to rainwater management in Rotterdam. Source: *Rotterdam Waterstad...* 2005, p. 75 171

Fig. 4.65: Benthemplein water square (project by De Urbanisten, 2013). On the left: a bird's-eye view (Rotterdam googlemaps). In the middle and on the right: a view of the square (photo by S. Gajek) 172

Fig. 4.66: London. On the left: city plan from the second half of the 17th century. On the right: unrealised plan of London development after the fire in 1666 by Ch. Wren (RIBA Collections) 174

Fig. 4.67: The subterranean rivers of London. Source: The Open Guide to London - london.openguides.org *https://commons.wikimedia.org/wiki/File:London_underground_rivers.jpg>* [accessed: 30.07.2018] 176

Fig. 4.68: London, view of the Thames and the vast post-industrial areas (photo by A. Januchta-Szostak). On the left: panorama of the city centre with "the O2" Millenium Dome 177

Fig. 4.69: London, St. Katharine Docks (photo by A. Januchta-Szostak) 177

List of figures 221

Fig. 4.70: London, Canary Wharf (photo by A. Januchta-Szostak). On the left: view from the waterfront towards London city centre panorama, on the right: ongoing process of the extension and modernisation of the docks (2015) 179

Fig. 4.71: London, ecological housing estates of *Greenwich Village located in the vicinity of Greenwich Ecology Park* (photos by A. Januchta-Szostak) 179

Fig. 4.72: London, Canada Waters (photos by A. Januchta-Szostak). Water purity is the condition for attractiveness of the former docks as the living environment. 180

Fig. 4.73: London, on the left: *Thames Barrier Park*, on the right: Thames Barrier – a big dam protecting the city from storm floods (photos by A. Januchta-Szostak) 180

Fig. 4.74: London, new bridges on the Thames – Millennium Bridge (photo by A. Januchta-Szostak) 181

Fig. 4.75: The Thames – landscape axis of London. Bridges' "gates" visible from the waterway. In the foreground: pedestrian Millennium Bridge (photo by A. Januchta-Szostak) 181

Fig. 4.76: London, the old and new icons of the Thames waterfront (photo by A. Januchta-Szostak). On the left: Tower Bridge and steep, inaccessible river banks. On the right: panorama of the left bank with the town hall (project by N. Foster) and the Shard skyscraper (project by R. Piano) 181

Fig. 4.77: London, the Barbican centre (photo by A. Januchta-Szostak, sat photo: Google maps) 182

Fig. 4.78: Development of the Lee river orientated towards enhancement of waterfront accessibility and cultural identity of the place (photo by A. Januchta-Szostak) 183

Fig. 4.79: Naturalised sections of the Lee river – reed biotopes facilitate water purification processes (photo by A. Januchta-Szostak) 184

Fig. 4.80: *The Blue Ribbon Network* (BRN) in London, source: ://www.london.gov.uk/what-we-do/planning/london-plan/current-london-plan/london-plan-chapter-seven-londons-living-space-4> [accessed: 7.09.2018] 185

Fig. 4.81: The current (2018) spatial development plan of Singapore (Urban Redevelopment Authority, *https://www.ura.gov.sg/maps2/?service=mp*> [accessed: 18.11.2018]) 188

Fig. 4.82: Keppel Container Terminal in Singapore (photo by Kroisenbrunner, *https://en.wikipedia.org/wiki/Port_of_Singapore*> accessed: [18.11.2018]) 189

Fig. 4.83: Singapore, on the left: Marina Bay Sands and Art Science Museum, on the right: Esplanade – Theatres on the Bay (photo by: Marcin Konsek, *https://pl.wikipedia.org/wiki/ Singapur*> [accessed: 18.11.2018]) 190

Fig. 4.84: Singapore, the view of the Marina Bay in the city centre (photo by: chensiyuan, 2012, *https://en.wikipedia.org/wiki/ Marina_Bay,_Singapore#/media/File:1_singapore_flyer_ view_2012.jpg*> [accessed: 18.11.2018]) 190

Fig. 4.85: Singapore, Bay South Garden (photo: *CC*, > [accessed: 18.11.2018]) 195

Fig. 4.86: Singapore. Buildings and plant structures of the *Gardens by the Bay* promote sustainable and energy-saving technological systems, allowing to minimise ecological footprint (source: Grant Associates,*http://grant-associates.uk.com/ projects/gardens-by-the-bay/*>[accessed: 18.11.2018]) 197

Fig. 4.87: Singapore, Kallang River @ Bishan-Ang Mo Kio Park (project by Atelier Dreiseitl, 2012). On the left: the valley with a concrete riverbed of the Kallang River. On the right: the river transformed into a meandering watercourse, source: *https:// howlingpixel.com/i-en/Bishan-Ang_Mo_Kio_Park*>) 197

Fig. 5.88: A diagram of spatial relations between a city and the environment as well as ratio of RGB structures (pie chart) in the RESPECT period (developed by A. Januchta-Szostak) 200

Fig. 5.89: A diagram of spatial relations between a city and the environment as well as ratio of RGB structures (pie chart) in the CONQUEST period (developed by A. Januchta-Szostak) 202

Fig. 5.90: A comparative profile: diagrams of changes in river valleys intersection and ratio of RGB structures in the urbanisation process (developed by A. Januchta-Szostak) 203

Fig. 5.91: A diagram of spatial relations between a city and the environment as well as the ratio of RGB structures (pie chart) in the RETURN period (developed by A. Januchta-Szostak) 204

Fig. 5.92: Revitalising rivers requires actions in valleys, tributaries systems and the entire catchment area, in all RGB structures (developed by A. Januchta-Szostak) 207

Fig. 5.93: Phases of changes in relations between a city and water, determined by key development factors (developed by A. Januchta-Szostak based on Brown et al. 2009) 209

List of tables

Tab. 2.1: The conquest period – the most critical changes within the river valleys and city catchment areas as well as their consequences - analysis of RGB urban structures. 84

Tab. 3.2: The profile of the most important international documents and events which influenced the change in city-river relation in terms of sustainable development as well as the enhancement of rivers and cities quality. 91

Tab. 3.3: Environmental ethics and various approaches to the relation with the environment (developed by: Świątek 2017 based on Thompson 2000). 99

Tab. 5.4: The template of cultural (R), environmental (G) and hydrological (B) goals orientated towards invigorating urban rivers (developed by A. Januchta-Szostak). 208

Bibliography

Ackroyd P., 2008, *Thames: Sacred River*, Vintage, London.
Ackroyd P., 2011, *London Under*, Chatto & Windus, London.
Arnold D., 1992, *Die Tempel Ägyptens, Götterwohnungen, Kultstätten*, Baudenkmäler, Zürich.
Bahri A., 2015, *Integrated urban water management*, Global Water Partnership, Stockholm, DOI: 10.13140/RG.2.1.4187.0160, <https://www.researchgate.net/publication/285729512_Integrated_urban_water_management> [Accessed: 22.02.2020].
Bajkiewicz-Grabowska E., Mikulski Z., 1999, *Hydrologia ogólna*, Wydawnictwo Naukowe PWN, Warszawa.
Balicki J., Bogucka M., 1989, *Historia Holandii* (2nd eddition) ., Zakład Narodowy im. Ossolińskich, Wrocław.
Bańkowska A., Sawa K., Popek Z., Wasilewicz M., Żelazo J., 2010, *Studia wybranych przykładów renaturyzacji rzek*, Infrastruktura i Ekologia Terenów Wiejskich nr 9, Komisja Techniczna Infrastruktury Wsi, PAN Oddział w Krakowie, p. 181–196.
Baranowski J.T., 1915, *Wsie holenderskie na ziemiach polskich*, Przegląd Historyczny, Vol. 19, Warszawa, p. 65–82.
Barker R., Coutts R., 2009, *Sustainable Development in Flood-risk Environments, Water Resource and Threat*, The International Review of Landscape Architecture and Urban Design, 68, p. 53–59.
Bartkowicz B., 1985, *Wpływ funkcji wypoczynku na kształtowanie struktury prze-strzennej miast*, Politechnika Krakowska, Kraków.
Bączkowska M., 2011, *"Artis sola dominanecessitas" – Otto Wagner's plans for the redevelopment of Vienna and Wienfluss*, [in:] A. Januchta-Szostak (ed.), *Sensitive approach to water in urban environment*, Series: Water in the Townscape, Vol. 4, Wydawnictwo Politechniki Poznańskiej, Poznań, p. 69–82.
Beatley T., 2000, *Green Urbanism: Learning From European Cities*, Island Press, Washington.
Benevolo L., 1980, *The History of the City*, MA: MIT Press, Cambridge, Massachusetts.
Benevolo L., 1995, *Miasto w dziejach Europy*, Seria: Tworzenie Europy, ed. J. Le Goff, Wydawnictwo Krąg, OW Wolumen, Warszawa.

Bergen S.C., Bolton S.M., Fridley J.L., 2001, *Design principles for ecological engineering*, Ecological Engineering, 18, Elsevier, p. 201–210.

Bernat S., 2007, *Rewitalizacja dolin rzecznych w miastach*, [in:] U. Myga-Piątek (ed.), *Doliny rzeczne. Przyroda – Krajobraz – Człowiek*, No. 7, Komisja Krajobrazu Kulturowego PTG, Oddział Katowicki PTG, Sosnowiec, p. 255–265.

Bernstein H.M., 2011, The Green Outlook 2011: Green Trends Driving Growth through 2015. F.ASCE, LEED AP, <http://www.nethyclan.com/Docs/BernsteinH2011GreenOutlookPresenttion.pdf> [Accessed: 24.02.2020].

Birkeland J., 2008, *Positive Development: From Vicious Circles to Virtuous Cycles through Built Environment Design*, Earthscan, London.

Biswas A.K., 1978, *Historia hydrologii*, PWN, Warszawa.

Blackbourn D., 2006, *The Conquest of Nature: Water, Landscape, and the Making of Modern Germany*, W.W. Norton & Company, New York and London, Reprint edition (August 17, 2007).

Bobiński E., Żelaziński J., 1996, *Czy można przerwać błędne koło ochrony przeciwpowodziowej?*, Gospodarka Wodna, No. 4, p. 99–107.

Bobiński E., Żelaziński J., 2015, *Ocena przyczyn lipcowej powodzi. Wnioski do programu ochrony przeciwpowodziowej w przyszłości na Odrze*, Ekspertyza opracowana dla Sejmowej Komisji Ochrony Środowiska, Zasobów Naturalnych i Leśnictwa, Warszawa, 15 września 1997, <www.odra.pl/pl/dokumenty/962585850.shtml> [Accessed: 16.07.2018].

Boeminghaus D., 1980, *Wasser im Stadtbild. Brunnen, Objekte, Anlagen*, Callwey, München.

Boer F., Jorritsma J., van Peijpe D., 2010, *De urbanisten en het wonder waterplein*, Uitgeverij 010, Rotterdam.

Bogdanowski J., 2000, *Rzeki i fortyfikacje*, [in:] J. Kołtuniak (ed.), *Rzeki. Kultura, cywilizacja, historia*, Vol. 9, Wydawnictwo Śląsk, Katowice, p. 203–215.

Bogucka M., 2011, *Z dziejów stosunków polsko-holenderskich w XVI–XVII wieku*. [From the history of Polish-Dutch relations in the 16th–17th centuries]. Czasy nowożytne, issue 24, p. 61–75.

Böhm A., 1994, *Architektura krajobrazu – jej początki i rozwój*, Wydawnictwo Politechniki Krakowskiej, Kraków.

Böhm A., 2006, *Planowanie przestrzenne dla architektów krajobrazu. O czynniku kompozycji*, Wydawnictwo Politechniki Krakowskiej, Kraków.

Bonenberg A., 2007, *Miasta i mosty*, Wydział Architektury Politechniki Poznańskiej, Poznań.

Bonenberg W., 2009, *Zaopatrzenie w wodę w cywilizacjach starożytnych*, Technologia Wody, 1(01), p. 7–20.

Boyd D., 2017, *The Rights of Nature: A Legal Revolution That Could Save the World*, ECW Press, Toronto.

Breen A., Rigby D., 1994, *Waterfronts. Cities Reclaim Their Edge*, Thames and Hudson, London.

Breen A., Rigby D., 1996, *The new Waterfront. A worldwide urban success story*, Thames and Hudson, London.

Broński K., Szpak J., 2002, *Procesy urbanizacyjne w Europie w XIX i XX wieku. Problemy i koncepcje badawcze*, Zeszyty Naukowe Akademii Ekonomicznej w Krakowie No. 587, p. 17–30, <https://r.uek.krakow.pl/bitstream/123456789/1194/1/13977.pdf> [Accessed: 22.02.2020].

Brown D.J., 2005, *Mosty. Trzy tysiące lat zmagań z naturą*, Wydawnictwo Arkady, Warszawa.

Brown L., 1996, *Bridges. Masterpieces of Architecture*, Todtri, London.

Brown P.H., 2009, *America's Waterfront Revival: Port Authorities and Urban Redevelopment*, University of Pennsylvania Press, Philadelphia.

Brown R., Keath N., Wong T., 2009, *Urban water management in cities: historical, current and future regimes*, Water Science and Technology, 59(5), p. 847–855.

Bruttomesso R. (ed.), 1993, *Waterfronts. A New Frontier for Cities on Water*, Centro Internacionale "CittaD'Acqua", Venice.

Bruttomesso R. (ed.), 1999, *Water and Industrial Heritage*, Marsilio Editori, Venice.

Bruttomesso R., 2011, *Complexity on the urban waterfront. Waterfronts in Post-Industrial Cities*, Spon Press, Londyn.

Bruttomesso R., Alemany J. (eds.), 2008, *Ciudades de Agua/Cities of Water/Cités d'eau*, Expoagua Zaragoza S.A./Prensa Diaria Aragonesa S.A., Zaragoza.

Burkiewicz Ł., Duchliński P., Kucharski J. (eds), 2014, *Oblicza wody w kulturze* [Faces of water in culture], HUMANITAS Studia Kulturoznawcze, Wydawnictwo Ignatianum, Wydawnictwo WAM, Kraków.

Byczkowski A., 1998, *Hydrologia*, Wydawnictwo SGGW, Warszawa.

Carpenter S.R., Pingali P.L., Bennett E.M., Zurek M.B. (eds.), 2005, *Ecosystems and human well-being: scenarios*, Island Press, Washington, D.C.

Carson R., 1962, *Silent Spring*, Houghton Mifflin Company, New York.

Chmielewski J.M., 1996, *Teoria urbanistyki. Wybrane zagadnienia*, Oficyna Wydawnicza Politechniki Warszawskiej, Warszawa.

Chopra K., Leemans R., Kumar P., Simons H. (eds.), 2005, *Ecosystems and human well-being: policy responses*, Island Press, Washington, D.C.

Chrzanowski T., 1995, *Geografia niderlandyzmu polskiego (XV–XVII wiek)*, [in:] T. Hrankowska (ed.), *Niderlandyzm w sztuce polskiej*, Materiały Sesji Stowarzyszenia Historyków Sztuki, Toruń, grudzień 1992, Warszawa, p. 59–80.

Cioc M., 2002, *The Rhine: An Eco-biography, 1815–2000*, University of Washington Press, Washington.

Clayton A., 2010, *Subterranean City: Beneath the streets of London*, Historical Publications, Whitstable, Kent.

Clouston B., Beaumont R., Rourke A., 1979, *The development of river valley landscapes in British cities*, [in:] I.C. Laurie (ed.), *Nature in Cities*, John Wiley & Sons, Chichester, p. 367–391.

Cole R.J., 2012, *Regenerative design and development: current theory and practice*, Building Research & Information, 40(1), Taylor & Francis Group, Routlege, p. 1–6, DOI: 10.1080/09613218.2012.617516.

Costanza R., d'Arge R., de Groot R., Farber S., Grasso M., Hannon B., Limburg K., Naeem S., Oneill R.V., Paruelo J., Raskin R.G., Sutton P., Van Den Belt M., 1997/5, *The value of the world's ecosystem services and natural capital*, Nature, 387, p. 253–260.

Crossman R., 2013, *Hunting for the Lost River of Paris*, <https://www.messynessychic.com/2013/12/09/hunting-the-lost-river-of-paris/> [Accessed: 29.02.2020].

Crouch D.P., 1993, *Water Management in Ancient Greek Cities*, Oxford University Press, New York.

Daily G., 1997, *Nature's Services: Societal Dependence on Natural Ecosystems*, Island Press, Washington, D.C.

Davidson N.C., 2014, *How much wetland has the world lost? Long-term and recent trends in global wetland area*, Marine and Freshwater Research, 65(10), p. 934–941. <https://continuite-ecologique.fr/wp-content/uploads/2017/12/ZH_Nick-C.-Davidson-2014.pdf> [Accessed: 29.02.2020]

DEFRA, 2005, *Making space for water*, Department for Environment, Food and Rural Affairs, London.

De Meulder B., Shannon K. (eds.), 2014, *UFO3 Water Urbanisms East*, Park Books, Zürich.

van Dijk G.M., Marteijn E.C.L., Schulte-Wulwer-Leidig A., 1995, *Ecological rehabilitation of the River Rhine: plans, progress and perspectives*, Regulated Rivers: Research & Management, 11, p. 377–388.

Dodge Data & Analytics, 2018, *World Green Building Trends 2018*, < <https://www.worldgbc.org/sites/default/files/World%20Green%20Building%20Trends%202018%20SMR%20FINAL%2010-11.pdf> [Accessed: 29.02.2020]

Drapella-Hermansdorfer A. (ed.), 2004, *Kształtowanie krajobrazu: idee, strategie, realizacje*, cz. 1. *Saksonia, Brandenburgia, Berlin*, Oficyna Wydawnicza Politechniki Wrocławskiej, Wrocław.

Drapella-Hermansdorfer A. (ed.), 2005, *Kształtowanie krajobrazu: idee, strategie, realizacje*, cz. 2. *Londyn i okolice*, Oficyna Wydawnicza Politechniki Wrocławskiej, Wrocław.

Drapella-Hermansdorfer A., 2005, *Zrównoważone miasta: Londyn, Paryż, Berlin między intencjami a realizacją*, [in:] A. Drapella-Hermansdorfer (ed.), *Oblicza równowagi – architektura, urbanistyka, planowanie – u progu Międzynarodowej Dekady Edukacji na rzecz Zrównoważonego Rozwoju*, Oficyna Wydawnicza Politechniki Wrocławskiej, Wrocław, pp. 74–82.

Drapella-Hermansdorfer A. (ed.), 2009, *Kształtowanie krajobrazu: idee, strategie, realizacje*, cz. 4. *Miasta nad wodą. Malmö, Kopenhaga, Hamburg*, Oficyna Wydawnicza Politechniki Wrocławskiej, Wrocław.

Directive 2000/60/EC of the European Parliament and of the Council of 23 October 2000 establishing a framework for Community action in the field of water policy (the so-called Water Framework Directive). <https://eur-lex.europa.eu/legal-content/EN/TXT/?qid=1582283380136&uri=CELEX:32000L0060> [Accessed: 21.02.2020].

Directive 2001/42/EC of the European Parliament and of the Council of 27 June 2001 on the assessment of the effects of certain plans and programmes on the environment. <https://eur-lex.europa.eu/legal-content/EN/TXT/?qid=1582283020635&uri=CELEX:32001L0042> [Accessed: 21.02.2020].

Directive 2007/60/EC of the European Parliament and of the Council of 23 October 2007 on the assessment and management of flood risks (the so-called Floods Directive). <https://eur-lex.europa.eu/legal-content/EN/TXT/?qid=1582282901343&uri=CELEX:32007L0060> [Accessed: 21.02.2020].

Directive 2004/35/CE of the European Parliament and of the Council of 21 April 2004 on environmental liability with regard to the prevention and remedying of environmental damage. <https://eur-lex.europa.eu/legal-content/EN/TXT/?qid=1582386678023&uri=CELEX:32004L0035> [Accessed: 22.02.2020].

Directive 2009/147/EC of the European Parliament and of the Council of 30 November 2009 on the conservation of wild birds. <https://eur-lex.europa.eu/legal-content/EN/TXT/?qid=1582386566798&uri=CELEX:32009L0147> [Accessed: 22.02.2020].

Council Directive 79/409/EEC of 2 April 1979 on the conservation of wild birds. <https://eur-lex.europa.eu/legal-content/EN/TXT/?qid=1582282731033&uri=CELEX:31979L0409> [Accessed: 21.02.2020].

Council Directive 85/337/EEC of 27 June 1985 on the assessment of the effects of certain public and private projects on the environment. <https://eur-lex.europa.eu/legal-content/EN/ALL/?uri=CELEX%3A31985L0337> [Accessed: 22.02.2020].

Council Directive 91/676/EEC of 12 December 1991 concerning the protection of waters against pollution caused by nitrates from agricultural sources.<https://eur-lex.europa.eu/legal-content/EN/TXT/?qid=1582282599730&uri=CELEX:31991L0676> [Accessed: 21.02.2020].

Council Directive 91/271/EEC of 21 May 1991 concerning urban waste-water treatment. <https://eur-lex.europa.eu/legal-content/EN/TXT/?qid=1582282241042&uri=CELEX:31991L0271> [Accessed: 21.02.2020].

Council Directive 92/43/EEC of 21 May 1992 on the conservation of natural habitats and of wild fauna and flora. <https://eur-lex.europa.eu/legal-content/EN/TXT/?uri=CELEX%3A31992L0043> [Accessed: 21.02.2020].

Edwards B., 1992, *London Docklands. Urban Design in the Age of Deregulation*, Butterworth Architecture, Oxford.

EC, 2013, The EU Strategy on adaptation to climate change, document adopted by the European Commission in April 2013, <https://ec.europa.eu/clima/sites/clima/files/docs/eu_strategy_en.pdf> [Accessed: 22.02.20120].

EEA, 2010, Environmental agreements since 1900, <https://www.eea.europa.eu/data-and-maps/figures/environmental-agreements-since-1900> [Accessed: 29.02.2020].

EEA, 2016a, Rivers and lakes in European cities. Past and future challenges, EEA Report No. 26/2016, Publication Office of the European Union, Luxembourg.

EEA, 2016b, Urban adaptation to climate change in Europe 2016. Transforming cities in a changing climate, Report No. 12/2016, Publications Office of the European Union, Luxembourg, <https://www.eea.europa.eu/publications/urban-adaptation-2016> [Accessed: 29.02.2020].

EEA, 2018, European waters. Assessment of status and pressures 2018, EEA Report No. 7/2018, Publication Office of the European Union, Luxembourg.

ENVIEDAN, 2013, Dealing with fluvial dynamics: A long-term interdisciplinary study of Vienna and the Danube, Water Histtory, secial issue 5/ 2013, p. 101–119, DOI 10.1007/s12685-013-0079-x <https://www.danubefuture.eu/sites/default/files/project/ENVIEDAN%20Special%20Issue%20Water%20History%202013.pdf> [Accessed: 3.08.2018].

Environment Agency, 2009, Thames Catchment Flood Management Plan, <https://www.gov.uk/government/publications/thames-catchment-flood-management-plan> [Accessed: 29.02.2020].

European Commission, 2013, Building a Green Infrastructure for Europe, EU, Belgium, <https://ec.europa.eu/environment/nature/ecosystems/docs/green_infrastructure_broc.pdf> [Accessed: 29.02.2020].

European Water Charter, 1968, EuropeanWater Charter.EN.txt, <https://iea.uoregon.edu/treaty-text/1968-europeanwatercharterentxt> [Accessed: 29.02.2020].

Feyen J., Shannon K., Neville M. (eds.), 2009, *Water & Urban Development Paradigms. Toward an integration of engineering, design and management approaches*, Proceeding of the International Urban Water Conference, Heverlee, Belgium, 15–19 September 2008, CRC Press Taylor & Francis Group, A Balkema Book, London.

Fisher B., Benson B., 2004, *Remaking the Urban Waterfront*, Urban Land Institute, Washington.

Flaga K., Januszkiewicz K., Hrabiec A., Cichy-Pazder E., 2005, *Estetyka konstrukcji mostowych*, Wydawnictwo Politechniki Krakowskiej, Kraków.

Forman R.T.T., Godron M., 1986, *Landscape ecology*, John Wiley and Sons, New York, NY, s. 619.

France R.L. (ed.), 2002, *Handbook of Water Sensitive Planning and Design* (Integra-ted Studies in Water Management and Land Development), Lewis Publishers, CRC Press, Boca Raton.

Freriks Ph., 2007, *Południk Paryża. Fascynująca wędrówka przez stolicę Francji*, transl. E. Zaleska, Wydawnictwo Trio, Warszawa.

Gan J.W., 1978, *Z dziejów żeglugi śródlądowej w Polsce*, Książka i Wiedza, Warszawa.

Garbrecht G., 1986, *Wasserspeicher (Talsperren) in der Antike*, Antike Welt, 2nd special edition, Antiker Wasserbau, (52), p. 51–64.

Garnier T., 1989, *An Industrial City*, Princeton Architectural Press, New York.

Gehl J., 1987, *The Life Between Buildings*, Van Nostrand Reinhold, New York. [Polish edition: *Życie między budynkami. Użytkowanie przestrzeni publicznych*, RAM, Kraków 2009.]

Geiger W., Dreiseitl H., 1999, *Nowe sposoby odprowadzania wód deszczowych*, Oficyna Wydawnicza Projprzem-Eko, Bydgoszcz.

Global Footprint Network, 2018, *National Footprint Accounts*, Global Footprint Network, Oakland, CA, USA.

Global Water Partnership (GWP), 2013, Integrated Urban Water Management (IUWM): Toward Diversification and Sustainability, Policy Brief, Global Water Partnership, Stockholm, <https://www.gwp.org/globalassets/global/toolbox/publications/policy-briefs/13-integrated-urban-water-management-iuwm.-toward-diversification-and-sustainability.pdf> [Accessed: 20.02.2020].

Gööck R., 1994, *Cuda świata*, 3 ed., Wydawnictwo Muza, Warszawa.

Graf R., Pyszny K., 2016, *Zintegrowana gospodarka wodna na obszarze metropolitalnym*, [in:] Ł. Mikuła (ed.), *Integracja planowania przestrzennego w metropolii Poznań – problemy, metody, osiągnięcia*, Centrum Badań Metropolitalnych UAM, Bogucki Wydawnictwo Naukowe, Poznań, p. 45–64.

Gray M.W., 2014, *Urban Sewage and Green Meadows: Berlin's Expansion to the South 1870–1920*, Central European History, 47, p. 275–306.

Green Capital. Green Infrastructure for a future ciy, 2016, Cross River Partnership, London, <https://www.london.gov.uk/sites/default/files/green_capital.pdf> [Accessed: 22.02.2020].

Green Infrastructure (GI) – Enhancing Europe's Natural Capital, 2013, Communication from the Commission to the European Parliament, the Council, the European Economic and Social Committee and the Committee of the Regions, Brussels, 6.5.2013 COM(2013) 249 final, <https://ec.europa.eu/environment/nature/ecosystems/docs/green_infrastructures/1_EN_ACT_part1_v5.pdf> [Accessed: 22.02.2020].

Green Paper on the Urban Environment, 1990, Commission of the European Communities, EC, COM (90) 218 final, 27 June 1990, Brussels, <http://aei.pitt.edu/1205/> [Accessed: 29.02.2020].

GUS, 2005, Ochrona środowiska. Informacje i opracowania statystyczne, GUS, Warszawa.

GUS, 2017, Ochrona środowiska. Informacje i opracowania statystyczne, GUS, Departament Badań Regionalnych i Środowiska, Warszawa.

Halliday S., 2013, *The Great Stink of London: Sir Joseph Bazalgette and the Cleansing of the Victorian Metropolis*, History Press Limited, Stroud.

Hassan R., Scholes R., Ash N. (eds.), 2005, *Ecosystems and human wellbeing: current state and trends*, Island Press, Washington.

Hausner J., 2017, *Ekonomia wartości a wartość ekonomiczna*, [in:] Open Eyes Book 2, Fundacja Gospodarki i Administracji Publicznej, Kraków, p. 23–75.

Hausner J., Kudłacz M., 2017, *Miasto-Idea – jak zapewnić rozwojową okrężność*, [in:] Open Eyes Book 2, Fundacja Gospodarki i Administracji Publicznej, Kraków, p. 195–228.

Hlavinek P., Zelenakova M. (eds.), 2015, *Storm Water Management*, Springer, Cham.

Hobhouse P., 2005, *Historia ogrodów*, Arkady, Warszawa.

Hohensinner S., 2012, *The Struggle with the River: Vienna and the Danube from 1500 to the Present*, Environment & Society Portal, Arcadia No. 19, Rachel Carson Center for Environment and Society, <http://www.environmentandsociety.org/arcadia/struggle-river-vienna-and-danube-1500-present> [Accessed: 29.02.2020].

Hohensinner S., Sonnlechner Ch., Schmid M., Winiwarter V., 2013, *Two steps back, one step forward: reconstructing the dynamic Danube riverscape under human influence in Vienna*, Water History, 5, p. 121–143, DOI: 10.1007/s12685-013-0076-0, <https://www.danubefuture.eu/sites/default/files/project/ENVIEDAN%20Special%20Issue%20Water%20History%202013.pdf> [Accessed:].

Holling C.D., 1978, *Adaptive Environmental Assesment and Management*, John Wiley, Chichester.

Hölzer Ch., Hundt T., Lüke C., Hamm O.G. (eds.), 2008, *Riverscapes. Designing Urban Embankments*, Montag Stiftung Urbane Räume, Birkhäuser, Basel–Boston–Berlin.

Hooimeijer F., Meyer H., Nienhuis A., 2005, *Atlas of Dutch water cities*, Uitgeverij SUN, Amsterdam.

Hu S., Niu Z., Chen Y., Li L., Zhang H., 2017, *Global wetlands: Potential distribution, wetland loss, and status*, Science of the Total Environment, 586, p. 319–327.

Hudson B., 1996, *Cities on the Shore. The Urban Littoral Frontier*, Pinter, London.

Hynes H.B.N., 1970, *The ecology of running waters*, Univ. Toronto Press, Toronto.

Ibelings H., 2003, *Supermodernism. Architecture in the Age of Globalization*, NAI Publishers, Rotterdam.

ICPR, 2013, Nathalie Plum, *Measures for river restoration in the international Rhine river basin*, International Commission for the Protection of the Rhine, Coblence, <http://m.stowa.nl/Upload/PPT%20presentaties%20bijeenkomsten/platformbijeen komst%20pbrh%2014112013/P06_PBRH_REFORM_Plum_%2014Nov2013.pdf> [Accessed: 5.09.2018].

IMGW, 2012, T. Walczykiewicz, R. Konieczny, P. Madej, M. Siudak, R. Bogdańska-Warmuz, I. Biedroń, Plany zarządzania ryzykiem powodziowym w Polsce, Magdeburg, <https://docplayer.pl/1622489-Plany-zarzadzania-ryzykiem-powodziowym-w-polsce.html> [Accessed: 29.02.2020].

Integrated Water Resources Management (IWMR), 2000, Global Water Partnership, Technical Advisory Committee, Background No. 4, Stockholm, Sweden.

IPCC, 2018, Global Warming of 1.5°C. An IPCC Special Report on the impacts of global warming of 1.5°C above pre-industrial levels and related global greenhouse gas emission pathways, in the context of strengthening the global response to the threat of climate change, sustainable development, and efforts to eradicate poverty, <https://www.ipcc.ch/sr15/> [Accessed: 24.02.2020]

IPCC, 2019, Special Report on the Ocean and Cryosphere in a Changing Climate, <https://www.ipcc.ch/site/assets/uploads/sites/3/2019/11/03_SROCC_SPM_FINAL.pdf> [Accessed: 24.02.2020]

IWA, [n.d.], City Water Stories, Singapore, <https://iwa-network.org/wp-content/uploads/2016/12/IWA_City_Stories_Singapore.pdf> [Accessed: 25.02.2020].

IWA, 2016a, Principles for Water-Wise Cities, International Water Association, London, <https://iwa-network.org/wp-content/uploads/2016/08/IWA_Brochure_Water_Wise_Cities.pdf> [Accessed: 25.02.2020].

IWA, 2016b, Water Utility Pathways in a Circular Economy, International Water Association, London, <http://www.iwa-network.org/wp-content/uploads/2016/07/IWA_Circular_Economy_screen-1.pdf> [Accessed: 25.02.2020].

Jadach-Sepioło A., 2007, *Gentryfikacja miast*, Problemy Rozwoju Miast, 4(3), p. 66–79, <http://bazhum.muzhp.pl/media//files/Problemy_Rozwoju_Miast/Problemy_Rozwoju_Miast-r2007-t4-n3/Problemy_Rozwoju_Miast-r2007-t4-n3-s66-79/Problemy_Rozwoju_Miast-r2007-t4-n3-s66-79.pdf> [Accessed: 25.02.2020].

Januchta-Szostak A., 2008, *Kreowanie tożsamości na styku wody i miasta*, Urbanista, 8, p. 31–33.

Januchta-Szostak A., 2010, *Miasto w symbiozie z wodą* [Town and Water Symbiosis], Czasopismo Techniczne – Technical Transactions, nr 6-A/1, z. 14, rok 107, special issue: *Miasto oszczędne* [Economical city], Vol. 2. p. 95-102.

Januchta-Szostak A., 2011a, *Front wodny Poznania – Dolina Warty. Rewitalizacja związków z rzeką* [Poznań Waterfront – Warta Valley. Revitalisation of the relationship with the river], Wydawnictwo Politechniki Poznańskiej, Poznań.

Januchta-Szostak A., 2011b, *Woda w miejskiej przestrzeni publicznej. Modelowe formy zagospodarowania wód opadowych i powierzchniowych*

[*Water in Urban Public Space. Model Forms of Rainwater and Surface Water Management*], Series: Rozprawy nr 454, Wydawnictwo Politechniki Poznańskiej, Poznań.

Januchta-Szostak A., 2012, *Kształtowanie miast wobec zagrożeń powodziowych w XXI wieku. Rotterdam – wodne miasto* [Shaping cities in the face of the risk of flooding in 21st century. Rotterdam – water city], Czasopismo Techniczne, Architektura, issue 1-A/1, vol. 1/109, p. 301–308.

Januchta-Szostak A., 2013, *Multifunctional riverside buffer parks – the research on nature-urban revitalisssation of river valleys*, Journal of Sustainable Architecture and Civil Engineering, 4(5), p. 42–50.

Januchta-Szostak A., 2014a, *Frontem do rzeki. Współczesne tendencje w zagospodarowaniu frontów wodnych i dolin rzecznych w miastach*, [in:] D. Bartkowiak (ed.), *Warta*, Seria: Kronika Miasta Poznania, Wydawnictwo Poznańskie, Poznań, p. 144–168.

Januchta-Szostak A., 2014b, *Rola urbanistyki i architektury w gospodarowaniu wodą*, [in:] T. Bergier, J. Kronenberg, I. Wagner (ed.), *Woda w mieście. Usługi ekosystemów dla zrównoważonej gospodarki wodnej*, Wydawnictwo Fundacji Sendzimira, Kraków, p. 31–47.

Januchta-Szostak A., 2015–2018, *Pro-ecological shaping of public places and buildings (Proekologiczne kształtowanie miejsc publicznych I budynków)*, Research project No. 10/01/DSPB/0252 – stage I, 10/01DSPB/0260 – stage II, 10.01/DSPB/0267 – stage III. (unpublished typescript), IAiPP, Wydział Architektury Politechniki Poznańskiej, Poznań.

Januchta-Szostak A., 2016, *Koncepcja placów deszczowych w Parku Rataje w Poznaniu*, [in:] M. Kosmala (ed.), *Tereny zieleni wobec zmian klimatu*, Wydawnictwo PZIiTS, Toruń 2016, p. 103–115.

Januchta-Szostak A., 2017a, *Geneza i znaczenie miejskich fontann*, Przestrzeń Miejska, 8, p. 34–38.

Januchta-Szostak A., 2017b, *Podejście zlewniowe w urbanistyce jako narzędzie zapobiegania powodziom miejskim*, [in:] W. Buczkowski, A. Szymczak-Graczyk (eds.), *Rewitalizacja obszarów zurbanizowanych. Powodzie w miastach – przyczyny, skutki, zapobieganie*, Wydawnictwo Prodruk, Poznań, p. 55–70.

Januchta-Szostak A., 2017c, *Regeneracja dolin rzecznych w miastach*, [in:] A. Januchta-Szostak, M. Banach (ed.), *Regeneracja miasta*, Seria: Człowiek–Ekologia–Ar-chitektura, Vol. 3, Wydawnictwo Politechniki Poznańskiej, Poznań, p. 41–60.

Januchta-Szostak A., 2018, *Miasta przyjazne wodzie?*, Open Eses Book 3, Fundacja Gospodarki i Administracji Publicznej, Kraków 2018, p. 165–185.

Januchta-Szostak A., Biedermann A., 2014, *The impact of Great Cultural Project on the transformation of urbanwater-sidespaces* [Wpływ wielkich projektów kulturalnych na przekształcenia miejskich obszarów nadwodnych], Czasopismo Techniczne, Seria: Architektura 1-A, Wydawnictwo Politechniki Krakowskiej, Kraków, p. 69–87.

Januchta-Szostak A., Biskupski P., 2014, S*pecyfika krajobrazu doliny Warty w Poznaniu*, [in:] D. Bartkowiak (ed.), *Warta*, Seria: Kronika Miasta Poznania, Wydawnictwo Poznańskie, Poznań, p. 129–244.

Johnson P., 1995, *Historia Anglików*, Marabut, Gdańsk.

Joint Steering Committee for Water Sensitive Cities (JSCWSC), 2009, *Evaluating options for water sensitive urban design – a national guide*, Joint Steering Committee for Water Sensitive Cities, Canberra, <http://observatoriaigua.uib.es/repositori/suds_australia_options.pdf> [Accessed: 24.02.2020].

Jokiel P., 2011, *„Mokre" konflikty*, Pracownia Hydrologii i Gospodarki Wodnej Uniwersytetu Łódzkiego, <http://hydro.geo.uni.lodz.pl/index.php?page=mokre-konflikty> [Accessed: 29.02.2020].

Kaminski M., 2000, *Venice. Art and Architecture*, Konemann.

Kaniecki A., 2004, *Poznań. Dzieje miasta wodą pisane*, Wydawnictwo PTPN, Poznań.

Klaiber G., 1996, *Hochwasserschutz durch Auerenaturierung am Oberrhein – Das Integrierte Rheinprogramm*, Wasserwirtschaft, 86(7/8), p. 396–400.

Knepper Th.P. (ed.), 2006, *The Rhine. The handbook of Environmental Chemistry*, Springer Science & Business Media, Berlin–Heidelberg.

Konieczny R., [n.d.], *Problemy z ograniczaniem skutków powodzi*, Biuro ds. Współpracy z Samorządami, Instytut Meteorologii i Gospodarki Wodnej, <http://www.malaretencja.pl/images/publikacje/Prezentacja_ograniczanie_skutkw_powodzi.pdf> [Accessed: 29.02.2020].

Ramsar treaty, 1971, Convention on Wetlands of International Importance especially as Waterfowl Habitat. Ramsar, Iran, 2.2.1971 as amended by the Protocol of 3.12.1982 and the Amendments of 28.5.1987, <https://www.ramsar.org/sites/default/files/documents/library/original_1971_convention_e.pdf> [Accessed: 22.02.2020].

Kosiński W., 2009, *Water in the Townscape and Cityscape – the Great and Plural Factor*, [in:] A. Januchta-Szostak (ed.), *Water in the Townscape*, Vol. 2, Wydawnictwo Politechniki Poznańskiej, Poznań, p. 19–44.

Kosiński W., 2011, *Perpendicular urban waterfront design – water plays the first role in composition*, [in:] A. Januchta-Szostak (ed.), *Sensitive approach to water in urban environment*, Series: Woda w Krajobrazie Miasta [Water in the Townscape], Vol. 4, Wydawnictwo Politechniki Poznańskiej, Poznań, p. 41–68.

Kosmala M. (ed.), 2011, *Miasta wracają nad wodę*, Polskie Zrzeszenie Inżynierów i Techników Sanitarnych Oddział Toruń, Toruń.

Kostof S., 1991, *The City Shaped: Urban Patterns and Meanings Through History*, Bullfinch Press, New York.

Kostrzewska M., 2013, *Miasto europejskie na przestrzeni dziejów. Wybrane przykłady*, Seria: Miasto • Metropolia • Region, ed. P. Lorens, Politechnika Gdańska, Akapit-DTP, Gdańsk.

Kowalczak P., 2007, *Konflikty o wodę*, Wydawnictwo Kurpisz, Przeźmierowo.

Kowalczak P., 2011, *Wodne dylematy urbanizacji*, Wydawnictwo Poznańskie, Poznań.

Kowalczak P., 2015, *Zintegrowana gospodarka wodna na obszarach zurbanizowanych. Część pierwsza: Podstawy hydrologiczno-środowiskowe*, Wydawnictwo ProDRUK, Poznań.

Krauze K., Wagner I., 2014. *Woda w przestrzeni miejskiej a zintegrowane zarządzanie miastem*, [in:] T. Bergier, J. Kronenberg, I. Wagner (ed.), *Woda w mieście. Zrównoważony rozwój – zastosowania*, cz. 5, Fundacja Sendzimira, Kraków, p. 95–114.

Krauze K., Żelewski Ł., Włodarczyk R., 2010, *Rola zieleni miejskiej w mieście przyszłości – Błękitno-Zielona Sieć Łodzi*, Acta Universitatis Lodziensis, Folia Biologica et Oecologica: Supplementum, Wydawnictwo Uniwersytetu Łódzkiego, Łódź.

Krawczuk A., 1998, *Rzeczni bogowie starożytnych Greków i Rzymian*, [in:] J. Kołtuniak (ed.), *Rzeki. Kultura, cywilizacja, historia*, Vol. 7, Wydawnictwo Śląsk, Katowice, p. 95–106.

Krawczuk A., Ostrowski J.A., Kaczanowski P., Zemanek A., Wolska-Lenarczyk J., Źrałka J., 2005, *Wielka historia świata. Tom 3 – Świat okresu cywilizacji klasycznych*, Oficyna Wydawnicza FOGRA, Kraków.

Kronenberg J., Bergier T. (ed.), 2010, *Wyzwania zrównoważonego rozwoju w Polsce*, Centrum Rozwiązań Systemowych Wrocław, Wydawnictwo Fundacji Sendzimira, Kraków.

Kuijpers J.W.M., 1995, *Ecological restoration of the Rhine/Maas estuary*, Water Science and Technology, 31(8), p. 187–195.

Kuitert W. (ed.), 2008, *Transforming with water. IFLA 2008 – Proceedings of the 45th World Congress of the International Federation of Landscape Architect*, Blauw-druk/Techne Press, Wageningen.

Kundzewicz Z.W. (ed.), 2012, *Changes in Flood Risk in Europe*, CRC Press, London.

Kundzewicz Z.W., 2018, *Zmiana klimatu i jej skutki – możliwości przeciwdziałania i adaptacji*, [in:] Open Eyes Book 3, Fundacja Gospodarki i Administracji Publicznej, Kraków, p. 141–161.

Kundzewicz Z.W., Kowalczak P., 2011, *Urban water – global challenges*, [in:] A. Januchta-Szostak (ed.), *Sensitive approach to water in urban environment*, Series: Water in the Townscape, Vol. 4, Wydawnictwo Politechniki Poznańskiej, Poznań, p. 85–96.

Lange K., Nissen S. (eds.), 2012, *Urban rivers – vital spaces. Guide for urban river revitalization*, REURIS Project Team, Leipzig.

Lauriel C., 1979, *Nature in Cities: Natural Environment in the Design and Development of Urban Green Areas*, John Wiley & Sons, New York.

LeGates R.T., Stout F. (eds.), 1996, *The City Reader*, Urban Reader Series, Routledge, New York.

Leipzig Charter on Sustainable European Cities, 2007, <http://www.sarp.org.pl/pliki/karta_lipska_pl.pdf> [Accessed: 13.01.2019].

Lisowska E., 2004, *Ogrody ziemi i raju, umysłu i duszy: Persja*, [in:] L. Sosnowski, A.I. Wójcik (ed.), *Ogrody – zwierciadła kultury*, t. 1. *Wschód*, TAiWPN Universitas, Kraków.

Liszewski S., 1995, *Geografia miast nadrzecznych*, [in:] J. Kołtuniak (ed.), *Rzeki. Kultura, cywilizacja, historia*, Vol. 4, Wydawnictwo Śląsk, Katowice. p. 127-151.

Londong D., 1993, *Der große Umbau im Emschergebiet*, Wasser und Boden, 45(3). p. 144-147.

Londong D., 1994, *Arbeiten für einen ökologisch ausgerichteten Umgang mit Regenswasser*, Wortrag Essener Tagung 1994, Gewässerschutz-Wasser-Abwasser, Bd. 1, Aachen.

Lorens P., 2001, *Rewitalizacja frontów wodnych nadmorskich miast portowych*, Wydział Architektury Politechniki Gdańskiej, Gdańsk [PhD thesis].

Lorens P., 2007, *Young City in Gdańsk as the Case in Urban Waterfront Redevelopment*, [in:] L. Nyka (ed.), *Water for Urban Strategies*, Verlag der Bauhaus-Universitat, Weimar.p. 82-87.

Lyle J.T., 1994, *Regenerative Design for Sustainable Development*, John Wiley & Sons, New York.

Łagiewski M., 1993, *Rzeczne przeprawy*, [in:] J. Kołtuniak (ed.), *Rzeki. Kultura, cywilizacja, historia*, Vol. 2, Wydawnictwo Śląsk, Katowice, p. 169–188.

Majdecki L., 1981, *Historia ogrodów. Przemiany, formy i konserwacja*, 2 ed., PWN, Warszawa.

Summary: *Making Space, Sharing Space: Fifth National Policy Document on Spatial Planning. 2000 / 2020*, 2001, Netherlands. Ministerie van Volkshuisvesting, Ruimtelijke Ordening en Milieubeheer (Ministry of Housing, Spatial Planning and the Environment, Communications Directorate), Den Haag, NSPA.

Makowski J., 1997, *Sztuka obwałowania rzek*, [in:] J. Kołtuniak (ed.), *Rzeki. Kultura, cywilizacja, historia*, Vol. 6, Wydawnictwo Śląsk, Katowice, p. 197–225.

Margul T., 1995, *Święte rzeki świata*, [in:] J. Kołtuniak (ed.), *Rzeki. Kultura, cywilizacja, historia*, t. 4, Wydawnictwo Śląsk, Katowice, p. 55–73.

Marsalek J., Cisneros B.J., Karamouz M., Malmquist P.A., Goldenfum J.A., Chocat B., 2008, *Urban Water Cycle Processes and Interactions*, Taylor & Francis Group, CRC Press, London.

Marshall R. (ed.), 2001, *Waterfronts in Post-Industrial Cities*, Taylor & Francis, Spons Press, London–New York.

Mayor of London, 2004, London Plan, Spatial Development Strategy for Greater London <https://www.london.gov.uk/what-we-do/planning/london-plan/past-versions-and-alterations-london-plan/london-plan-2004> [Accessed: 29.02.2020].

Mayor of London, 2011, London Water Strategy (LWS), Securing London's water future. The Mayor's water Strategy. October 2011. Greater London Authority, London, <https://www.london.gov.uk/sites/default/files/gla_migrate_files_destination/water-strategy-oct11.pdf> [Accessed: 29.02.2020].

Mayor of London, 2018, London Environment Strategy (LES), Greater London Authority, <https://www.london.gov.uk/what-we-do/environment/london-environment-strategy> [Accessed: 20.02.2020].

McHarg J., 1969, *Design with Nature*, Doubleday/Natural History Press, New York.

Meyer H., 1999, *City and Port*, International Books, Rotterdam.

Mill J.S., 1965, *Zasady ekonomii politycznej*, PWN, Warszawa.

Millennium Ecosystem Assessment, 2005, Ecosystems and human well-being: synthesis, World Resources Institute, Island Press, Washington, DC, <http://www.millenniumassessment.org/documents/document.356.aspx.pdf> [Accessed: 29.02.2020].

Mini P.V., 1974, *Philosophy and Economics: The Origin and Development of Econo-mic Theory*, University Presses of Florida, Gainesville.

Mitchell R.B., 2017, *International Environmental Agreements Database Project* (Version 2017.1), University of Oregon, Eugene, OR, <https://iea.uoregon.edu/> [Accessed: 29.02.2020].

Montesquieu (Monteskiusz) Ch., 1927, *O duchu praw*, Wydawnictwo F. Hoesick, Warszawa, on-line version: Publisher: Fundacja Nowoczesna Polska, <https://wolnelektury.pl/media/book/pdf/o-duchu-praw.pdf> [Accessed: 22.02.2020].

Morris A.E.J., 1994, *History of Urban Form: Before the Industrial Revolutions*, Wiley, New York.

Mumford L., 1934, *Technics and Civilization*, Harcourt, Brace & Company, Inc., New York.

Mumford L., 1944, *Condition of Man*, Harcourt, Brace, New York.

Mumford L., 1961, *The City in History. Its Origins, Its Transformation and Its Properties*, Harcourt, Brce & World Inc, New York.

Nachlik E., 2006, *Ochrona przeciwpowodziowa w powiązaniu z ochroną walorów przyrodniczych rzek i ich dolin*, Infrastruktura i Ekologia Terenów Wiejskich, 4(1), Polska Akademia Nauk, Oddział w Krakowie, Komisja Technicznej Infrastruktury Wsi, p. 47–62.

Nagle G., 2003, *Access to Geography: Rivers & Water Management*, Arnold Publishers, London.

Nash R.F., 1989, *The Rights of Nature: A History of Environmental Ethics* (History of American

New Charter of Athens (Nowa Karta Ateńska), 1998, Zasady planowania miast przyjęte przez Europejską Radę Urbanistów, Biuletyn Informacyjny TUP, nr specjalny, Warszawa.

New Charter of Athens (Nowa Karta Ateńska), 2003, *Wizja miast XXI wieku*, Zasady planowania miast przyjęte przez Europejską Radę Urbanistów, przekł. z franc., Biuletyn Informacyjny TUP, nr specjalny, Warszawa.

Niemczyk E., 1995, *Motywy akwatyczne w architekturze*, [in:] J. Kołtuniak (ed.), *Rzeki. Kultura, cywilizacja, historia*, Vol. 4, Wydawnictwo Śląsk, Katowice, p. 75–126.

Niemczyk E., 2002, *Cztery żywioły w architekturze*, Ossolineum, Wrocław.

Ninck M., 1967, *Die Bedeutung des Wassers im Kult und Leben der Alten. Eine symbolgeschichtliche Untersuchung*, Wissenschaftliche Buchgesellschaft, Darmstadt.

Norgaard R.B., 2010, *Ecosystem services: From eye-opening metaphor to complexity blinder*, Ecological Economics, 69(6), Elsevier,

p. 1219–1227, <https://www.sciencedirect.com/science/article/abs/pii/S0921800909004583?via%3Dihub> [Accessed: 29.02.2020].

Nowakowski Z., 1931, *Geografia serdeczna*, GiW, Warszawa.

Nyka L., 2007, *Architecture and Water – New Concepts on Blurring Borders*, [in:] L. Nyka (ed.), *Water for urban strategies*, Verlag der Bauhaus-Universitat, Weimar. p. 20-27

Nyka L. (ed.), 2007, *Water for urban strategies*, Verlag der Bauhaus-Universitat, Weimar.

Nyka L., 2013, *Architektura i woda – przekraczanie granic*, Wydawnictwo Politechniki Gdańskiej, Gdańsk.

Oates R., 2012, *The London Rivers Action Plan. Thames River Restoration Trust*, 15th International River Symposium, Melbourne, 8–11.10.2012, <http://archive.river-symposium.com/index.php?element=W_s4_B3_Robert+Oates.pdf> [Accessed: 22.08.2018].

Ocioszyński T., 1968, *Rozwój żeglugi i myśli morskiej*, Wydawnictwo Morskie, Gdynia.

O'Donnell E., 2018, *Legal Rights for Rivers. Competition, Collaboration and Water Governance*, 1st Edition, Series: Earthscan Studies in Water Resource Management, Routledge, Abingdon, Oxon, New York, NY.

O'Donnell E.L., Talbot-Jones J., 2018, *Creating legal rights for rivers: lessons from Australia, New Zealand, and India*, Ecology and Society, 23(1), 7, <https://www.ecologyandsociety.org/vol23/iss1/art7/> [Accessed: 29.02.2020].

Odum E.P., 1953, *Fundamentals of ecology*, W.B. Saunders Company, Philadelphia.

Odum H., 1971, *Environment, power, and society*, Wiley-Interscience, London.

OECD, 2009, Declaration on Green Growth, Adopted at the Meeting of the Council at Ministerial Level on 25 June 2009 [C/MIN(2009)5/ADD1/FINAL], <https://www.oecd.org/env/44077822.pdf> [Accessed: 29.02.2020].

OECD, 2011, *Towards Green Growth*, OECD Green Growth Studies, OECD Publishing, Paris, <https://www.oecd-ilibrary.org/environment/towards-green-growth_9789264111318-en> [Accessed: 29.02.2020].

OECD, 2015, *OECD Principles on Water Governance*, <https://www.oecd.org/cfe/regional-policy/OECD-Principles-on-Water-Governance.pdf> [Accessed: 29.02.2020].

Orłowski B., 1993, *Rzeki jako wyzwanie dla techniki*, [in:] J. Kołtuniak (ed.), *Rzeki. Kultura, cywilizacja, historia*, Vol. 2, Wydawnictwo Śląsk, Katowice, p. 37–56.

Ostrowski W., 2001, *Wprowadzenie do historii budowy miast. Ludzie i środowisko*, Oficyna Wydawnicza Politechniki Warszawskiej, Warszawa.

Otto B., McCormick K., Leccese M., 2004, *Ecological Riverfront Design: Restoring Rivers, Connecting Communities*, American Planning Association, Planning Advisory Service, Report Number: 518–519, Chicago, IL.

Pancewicz A., 2004, *Rzeka w krajobrazie miasta*, Wydawnictwo Politechniki Śląskiej, Gliwice.

Pedersen Zari M., 2012, *Ecosystem services analysis for the design of regenerative built environments*, Building Research & Information, 40, 1, 1–6, p. 54–64.

Peterson G.D., Allen C.R., Holling C.S., 1998, *Ecological resilience, biodiversity and scale*, Ecosystem, 1, p. 6–18.

Petryshyn H., 2015, *Kulturowy waterfront Kopenhagi* [Cultural waterfront of the Copenhagen city], Space & FORM/przestrzeń i FORMA, 24(2), p. 95–112.

Petryshyn H., Sosnova N., 2016, *Prospekt Wolności punktem kulminacyjnym w poszukiwaniu samoidentyfikacji Lwowa*, [in:] A. Januchta-Szostak, M. Banach (ed.), *Zrównoważone miasto – idee i realia*, Seria: Człowiek–Ekologia–Architektura, Vol. 1, Wydawnictwo Politechniki Poznańskiej, Poznań, p. 111–124.

Pickett S.T.A., Burch W.R. Jr, Dalton S.E., Foresman T.W., Grove J.M., Rowntree R.A., 1997, *Conceptual framework for the study of human ecosystems in urban areas*, Urban Ecosystems, 1, p. 185–199.

Piskozub A., 1993, *Wielkie cywilizacje rzeczne*, [in:] J. Kołtuniak (ed.), *Rzeki. Kultura, cywilizacja, historia*, t. 2, Wydawnictwo Śląsk, Katowice, p. 11–35.

Planning Policy Statement 25: Development and Flood Risk Practice Guide, 2009, Department for Communities and Local Government, London.

Pötz H., Bleuzé P., 2016, *Urban green-blue grids for sustainable and dynamic cities*, coop for life, Delft.

PUB, 2014, Managing Stormwater for Our Future. Public Utilities Board, Singapore, <https://www.pub.gov.sg/Documents/ManagingStormwater.pdf > [Accessed: 25.02.2020]

PUB, 2018, Active, Beautiful, Clean Waters. Design Guidelines, 4th Edition. Public Utilities Board, Singapore, <https://www.pub.gov.sg/Documents/ABC_Waters_Design_Guidelines.pdf> [Accessed: 25.02.2020].

Rees W.E., 1992, *Ecological footprints and appropriated carrying capacity: what urban economics leaves out*, Environment and Urbanisation, 4(2), p. 121–130.

REURIS project team, 2012, *Urban Rivers – Vital Spaces. Manual for urban river revitalisation – implementation, participation, benefits*. Bydgoszcz, pp. 327, Projekt REURIS, Katowice–Stuttgart, <https://www.yumpu.com/en/document/read/25724375/manual-for-urban-river-revitalisation-central-europe> [Accessed: 29.02.2020].

RIBA, 2009, *Designing for Flood Risk*, 07 Climate Change Toolkit, <http://www.gbc.ee/757est.pdf> [Accessed: 20.02.2020].

River Restoration Centre (RRC), (n.d.), Manual of River Restoration Techniques, <https://www.therrc.co.uk/manual-river-restoration-techniques> [Accessed: 22.02.2020], Polish version: *Przyjazne naturze kształtowanie rzek i potoków – praktyczny podręcznik*, 2006, Polska Zielona Sieć, Wrocław–Kraków.

River Restoration Centre, 2009, London Rivers Action Plan (LRAP), <http://www.therrc.co.uk/lrap/lplan.pdf> [Accessed: 29.02.2020].

Rivers by Design, 2013, Rethinking development and river restoration. A guide for planners, developers, architects and landscape architects on maximizing the benefits of river restoration, Partners of the Restore Project, <https://assets.publishing.service.gov.uk/government/uploads/system/uploads/attachment_data/file/297315/LIT8146_7024a9.pdf> [Accessed: 20.02.2020].

Romanowicz R.J., Nachlik E., Januchta-Szostak A., Starkel L., Kundzewicz Z.W., Byczkowski A., Kowalczak P., Żelaziński J., Radczuk L., Kowalik P., Szamałek K., 2014, *Zagrożenia związane z nadmiarem wody*, Kwartalnik NAUKA, 1, p. 123–148.

Rotterdam Resilience Strategy, 2016, Ready For the 21st Century, <https://s3.eu-central-1.amazonaws.com/storage.resilientrotterdam.nl/uploads/2017/11/09115607/strategy-resilient-rotterdam.pdf> [Accessed: 29.02.2020].

Rotterdam Waterstad 2035, 2005, Internationale Architectuur Biënnale Rotterdam 2005, Episode Publishers, Rotterdam.

Ruano M., 1998, *Eco-Urbanism: Sustainable Human Settlements, 60 Case Studies*, Gustavo Gili, Barcelona.

Rusiński W., 1947, *Osady tzw. „Olędrów" w dawnym województwie poznańskim*, Polska Akademia Umiejętności, Kraków.

Schneider-Skalska G., 1997, *Tereny nadwodne jako obszary publiczne*, Zeszyty Naukowe IPU, nr 5, Politechnika Krakowska, Kraków.

Schultz G., 1995, *Informationen zur gegenwärtigen Hochwassersituation am Rhein*, Wasserwirtschaft, 85(4), p. 216.

Sedláček T., 2012, *Ekonomia dobra i zła. W poszukiwaniu istoty ekonomii od Gilgamesza do Wall Street*, Wydawnictwo Studio Emka, Warszawa.

Seltmann G., 2007, *Renaissance of an Industrial Region: "Internationale Bauausstellung Emscher Park" – achievements and future model for others*, Gse Project for regional development, Flechtingen, Germany, <http://www.riss.osaka-u.ac.jp/jp/events/point/P.Seltmann.pdf> [Accessed: 20.02.2020].

Seroka K., 2012, *Haussmann i przebudowa Paryża: od średniowiecznej zabudowy do nowoczesnej metropolii*, Portal historyczny: Histmag.org, licencja: CC BY-SA 3.0, <https://histmag.org/Georges-Haussmann-i-wielka-przebudowa-Paryza-od-sredniowiecznej-zabudowy-do-nowoczesnej-metropolii-6669> [Accessed: 29.02.2020].

Shannon K., De Meulder B., 2008, *Water and the City: the "Great Stink" and Clean Urbanism*, [in:] K. Shannon, B. De Meulder, J. Gosseye, V. D'Auria (eds.), *UFO1 Water Urbanisms*, SUN, Amsterdam, p. 5–9.

Shannon K., De Meulder B., Gosseye J., D'Auria V. (eds.), 2008, *UFO1 Water Urbanisms*, SUN, Amsterdam.

Sima U., Umweltstadträtin W., Loew G., 2017, *Donauhochwasserschutz Wien. Flood Control on the Danube, Vienna*, Leiter der Fachabteilung MA 45 – Wiener Gewässer, Wien, <https://www.wien.gv.at/umwelt/gewaesser/pdf/donau-hochwasserschutz-2017.pdf> [Accessed: 29.02.2020].

Słyś D., 2008, *Retencja i infiltracja wód deszczowych*, Oficyna Wydawnicza Politechniki Rzeszowskiej, Rzeszów.

Słyś D., 2013, *Zrównoważone systemy odwodnienia miast*, Dolnośląskie Wydawnictwo Edukacyjne, Wrocław.

Sosnowski L., Wójcik A.I. (eds.), 2004, *Ogrody – zwierciadła kultury*, t. 1. Wschód, TAiWPN Universitas, Kraków.

Sosnowski L., Wójcik A.I. (eds.), 2008, *Ogrody – zwierciadła kultury*, t. 2. Zachód, TAiWPN Universitas, Kraków.

SPA 2020, 2013, Strategiczny plan adaptacji dla sektorów i obszarów wrażliwych na zmiany klimatu do 2020 roku z perspektywą do roku 2030, Ministerstwo Ochrony Środowiska, Warszawa, <https://bip.mos.gov.pl/fileadmin/user_upload/bip/strategie_plany_programy/Strategiczny_plan_adaptacji_2020.pdf> [Accessed: 29.02.2020].

Stanielewicz J., 1995, *Rzeki jako warsztat pracy*, [in:] J. Kołtuniak (ed.), *Rzeki. Kultura, cywilizacja, historia*, t. 4, Wydawnictwo Śląsk, Katowice, p. 153–175.

Steffen W., Broadgate W., Deutsch L., Gaffney O., Ludwig C., 2015, *The trajectory of the Anthropocene: The Great Acceleration*, The Anthropocene Review, 2(1), p. 81–98, 2015-04-01, <https://journals.sagepub.com/doi/abs/10.1177/2053019614564785?journalCode=anra> [Accessed: 29.02.2020].

Stolińska M., 2018, *Znikające rzeki Łodzi*, Politechnika Poznańska, Poznań [study work under the supervision of dr hab. eng. arch. A. Januchta-Szostak as part of the Research Studio 2017/2018 at the Faculty of Architecture of the Poznań University of Technology (unpublished)].

Stoner A., Melathopoulos A., 2015, *Freedom in the Anthropocene: Twentieth-Century Helplessness in the Face of Climate Change*, Palgrave Macmillan, New York.

Sukopp H., Blume H.P., Kunick W., 1979, *The soil, flora and vegetation of Berlin's wastelands*, [in:] I.C. Laurie (ed.), *Nature in Cities*, John Wiley, Chichester, p. 115–132.

SWITCH – Managing Water for the City of the Future, 2006–2011, Findings from the SWITCH Project2006-2011, <Sustainable Water Management in the City of the Future> [Accessed: 29.02.2020].

Swyngedouw E., Kaika M., 2000, *The environment of the city or . . . the urbanisation of nature*, [in:] G. Bridge, S. Watson (eds.), *A Companion to the City*, Blackwell, Oxford, p. 567–580.

Szałkiewicz E., Jusik S., Grygoruk M., 2018, *Status of and Perspectives on River Restoration in Europe: 310,000 Euros per Hectare of Restored River*, Sustainability, 10(29), DOI:10.3390/su10010129, <https://www.preprints.org/manuscript/201712.0033/v1> [Accessed: 29.02.2020].

Szałygin J., [n.d.], *Osadnictwo olęderskie na środkowym Mazowszu*, Wirtualne Muzeum Konstancina, <file:///C:/Users/Użytkownik/AppData/Local/Temp/muzeumkonstancina.pl_20200229212845.pdf> [Accessed: 29.02.2020].

Szulczewska B., 2002, *Teoria ekosystemu w koncepcjach rozwoju miast*, Wydawnictwo SGGW, Warszawa.

Świątek L., 2017, *Architektura pozytywnego rozwoju*, [in:] A. Januchta-Szostak, M. Banach (ed.), *Regeneracja miasta*, Seria: Człowiek–Ekologia–Architektura, t. 3, Wydawnictwo Politechniki Poznańskiej, Poznań, p. 7–17.

Temple R., 1994, *Geniusz Chin*, Ars Polona, Warszawa.

Thompson I.H., 2000, *The Ethics of Sustainability*, Landscape and Sustainability, Spon Press, London.

Tjallingii S.P., 1995, *Ecopolis: strategies for ecologically sound urban development*, Backhuys Publishers, Leiden.

Tölle-Kastenbein R., 1990, *Antike Wasserkultur*, Beck, München.

Tvedt T., Jakobsson E. (eds.), 2006, *A History of Water*, Series I (3 vols.), I.B. Tauris, London.

UN, 2015, Transforming our World: The 2030 Agenda for Sustainable Development, A/RES/70/1, <https://sustainabledevelopment.un.org/content/documents/21252030%20Agenda%20for%20Sustainable%20Development%20web.pdf> [Accessed: 22.02.2020]

UN, 1972, Declaration of the United Nations Conference on the Human Environment, A/CONF.48/14/Rev. 1, Chapter 1, Stockholm, June 1972.

UN, 1992, Rio Declaration on Environment and Development, A/CONF.151/26 (Vol. 1), Chapter 1, Annex 1, Rio de Janeiro, June 1992.

UN, 2002a, Johannesburg Declaration on Sustainable Development, A/CONF. 199/20, Chapter 1, Resolution 1, Johannesburg, September.

UN, 2002b, Plan of Implementation of the World Summit on Sustainable Development, A/CONF.199/20, Chapter 1, Resolution 2, Johannesburg, September.

UN DESA, 2018, World Urbanization Prospects: The 2018 Revision, <https://www.un.org/development/desa/publications/2018-revision-of-world-urbanization-prospects.html> [Accessed: 29.02.2020].

UN-Habitat, 1996, An Urbanizing World – Global Report on Human Settlements 1996, Oxford University Press, Oxford.

UN-Habitat, 2009, Planning Sustainable Cities – Global Report on Human Settlements 2009, Series: Global Report on Human Settlements, <http://mirror.unhabitat. org/categories.asp?catid=555> [Accessed: 29.02.2020].

UN-Habitat, 2011, Cities and Climate Change: Global Report on Human Settlements 2011, Series: Global Report on Human Settlements, <http://mirror.unhabitat. org/categories.asp?catid=555> [Accessed: 29.02.2020].

UN-Habitat III, 2016, The New Urban Agenda, <http://habitat3.org/wp-content/uploads/NUA-English.pdf> [Accessed: 22.02.2020].

URBEM – Urban River Basin Enhancement Methods, 2004, Final Report. Leibniz Institute of Ecological and Regional Development, Dresden (IOER), Dresden University of Technology (TU Dresden).

Vassari G., 1998, *The Lives of the Artists* (Oxford World's Classics), Oxford University Press, Oxford.

de Waal L.C., Large A.R.G., Wade P.M., 1998, *Rehabilitation of rivers – Principles and Implementation*, J. Wiley and Sons, Chichester [reprint 2000].

Wackernagel M., Rees W., 1996, *Our Ecological Footprint, Reducing Human Impact on the Earth*, New Society Press, British Columbia.

Wagner I., Januchta-Szostak A., Waack-Zając A., 2014, *Narzędzia planowania i zarządzania strategicznego wodą w przestrzeni miejskiej*, [in:] T. Bergier, J. Kronenberg, I. Wagner (ed.), *Woda w mieście. Usługi ekosystemów dla zrównoważonej gospodarki wodnej*, Wydawnictwo Fundacji Sendzimira, Kraków, p. 17–29.

Wagner I., Krauze K., Zalewski M., 2013, *Błękitne aspekty zielonej infrastruktury*, [in:] T. Bergier, J. Kronenberg, P. Lisicki (ed.), *Przyroda w mieście – rozwiązania*, Seria: Zrównoważony Rozwój – Zastosowania nr 4, Wydawnictwo Fundacji Sendzimira, Kraków, p. 145–155.

Waldheim Ch. (ed.), 2006, *The Landscape Urbanism Reader*, Princeton Architectural Press, New York.

Walter B., 1975, *Twórca jako wytwórca*, Wydawnictwo Poznańskie, Poznań.

Water Environment Foundation (WEF), 2009, The Baltimore Charter for Sustainable Water Systems, Baltimore, DOI:10.2175/193864709793900311 <http://sustainablewaterforum.org/baltimore.html> [Accessed: 24.02.2020].

Water Spaces, 2006, Vol. 2. *A Pictorial Review*, Images Publishing Dist A/C, 08(10).

Waterplan 2, 2007, Working on water for an attractive city Rotterdam, Gemeente Rotterdam.

Watson D., Adams M., 2011, *Design for flooding: Architecture and Landscape Design for Resilience to Climate Change*, John Willey & Sons, New Jersey.

Wawręty R. (ed.), 2004, *Renaturyzacja rzek*, Towarzystwo na rzecz Ziemi, Oświęcim, <http://www.ratujmyrzeki.pl/dysk_KRR/biblioteka_koalicji/2004_5_Renaturyzacja_rzek_TnZ.pdf> [Accessed: 29.02.2020].

Wawręty R., Żelaziński J., 2007, *Środowiskowe skutki przedsięwzięć hydrotechnicznych współfinansowanych ze środków Unii Europejskiej*, Raport Towarzystwa na rzecz Ziemi i Polskiej Zielonej Sieci, Oświęcim–Kraków, <http://docplayer.pl/4709010-Srodowiskowe-skutki-przedsiewziec-hydrotechnicznych-wspolfinansowanych-ze-srodkow-unii-europejskiej.html> [Accessed: 29.02.2020].

Wines J., 2008, Seria: Zielona architektura, Ph. Jodidio (ed.), Taschen, Köln.

Winiwarter V., Schmid M., Dressel G., 2013, *Looking at half a millennium of co-existence: the Danube in Vienna as a socio-natural site*, Water History, 5(2), p. 101–119, <https://link.springer.com/article/10.1007/s12685-013-0079-x> [Accessed: 20.02.2020].

Vitruvius M.P. (Witruwiusz), 1999, *O architekturze ksiąg dziesięć*, Wydawnictwo Prószyński i S-ka, Warszawa.

Wittfogel K., 1957, *Oriental despotism: A comparative study o total power*, Yale University Press, New Haven.

Wojciechowski K.H., 2000, *Człowiek i rzeka w układzie przyrodniczym i gospodarczym*, [in:] J. Kołtuniak (ed.), *Rzeki. Kultura, cywilizacja, historia*, Vol. 9, Wydawnictwo Śląsk, Katowice, p. 187–201.

Wojnowska-Heciak M., 2017, *Tendencje rozwoju i kształtowania obszarów położonych nad rzekami w aglomeracjach miejskich*, Wydział Architektury

Politechniki Warszawskiej, Warszawa [PhD thesis supervised by dr hab. inż. arch. Joanna Giecewicz, prof. PW – unpublished.].

Word Commission on Environment and Development, 1987, Report of the World Commission on Environment and Development: Our Common Future (Brundtland Report), <http://www.un-documents.net/our-common-future.pdf> [Accessed: 29.02.2020].

World Economic Forum, 2018, The Global Risks Report 2018, 13th Edition, Geneva, <http://www3.weforum.org/docs/WEF_GRR18_Report.pdf> [Accessed: 29.02.2020].

Wrenn D.M., Casazza J.A., Smart J.E., 1983, *Urban Waterfront Development*, The Urban Land Institute, Washington.

WSUD, 2005, *Engineering Procedures: Stormwater*, Melbourne Water, Csiro Publishing, Australia.

WWAP (United Nations World Water Assessment Programme), 2018, The United Nations World Water Development Report 2018: Nature-based Solutions, Paris, UNESCO.

WWDP (United Nations World Water Development Report), 2018, UN World Water Development Report: Nature-based Solutions for Water. Executive Summary, <https://unesdoc.unesco.org/ark:/48223/pf0000261594> [Accessed: 20.02.2020].

Zachariasz A., 2006, *Zieleń jako współczesny czynnik miastotwórczy ze szczególnym uwzględnieniem roli parków miejskich*, Wydawnictwo Politechniki Krakowskiej, Kraków.

Zalewski M., 2011, *Ecohydrology for implementationof the UE water framework directive*, Proceedings of the Institution of Civil Engineering Water Management, 164, p. 375–385.

Zalewski M., 2014, *Woda jako podstawa jakości życia w miastach przyszłości*, [in:] T. Bergier, J. Kronenberg, I. Wagner (ed.), *Woda w mieście. Usługi ekosystemów dla zrównoważonej gospodarki wodnej*, Wydawnictwo Fundacji Sendzimira, Kraków, p. 9–15.

Zaraś-Januszkiewicz E.M., Fornal-Pieniak B., Żarska B., 2013, *Pozostałości osadnictwa holenderskiego jako element historycznej zielonej infrastruktury w krajobrazie kulturowym centralnej i północnej części Mazowsza [Remains of the Dutch colonization as an element of historical green infrastructure in cultural landscape of central and north part of Mazovia]* Problemy Ekologii Krajobrazu, Vol. 36, p. 95–108, <http://agro.icm.edu.pl/agro/element/bwmeta1.element.agro-ffe2d5c1-adb7-4c65-b5d5-151ca7873ede> [Accessed: 20.02.2020].

Zujewski B., 2014, *Burzliwe dzieje miejskiego strumyka*, Zieleń Miejska, 3, <http://e-czytelnia.abrys.pl/zielen-miejska/2014-3-743/projekty-i-realizacje-8675/burzliwe-dzieje-miejskiego-strumyka-17693> [Accessed: 29.02.2020].

Żelaziński J., [n.d.], *Techniczne środki ochrony przeciwpowodziowej i ich zawodność – przykłady polskie i zagraniczne*, IMGW, Warszawa, <http://www.tnz.most.org.pl/dokumenty/publ/psopp/imgw_w3.htm> [Accessed: 1.08.2018].

Żelazo J., 2006, *Renaturyzacja rzek i dolin*, Infrastruktura i Ekologia Terenów Wiejskich, 4(1), PAN, Oddział w Krakowie, Komisja Technicznej Infrastruktury Wsi, pp. 11–31.

Żelazo J., Popek Z., 2002, *Podstawy renaturyzacji rzek*, Wydawnictwo SGGW, Warszawa.

Żukow-Karczewski M., [n.d.], *Wodociągi. Historia wodociągów w Polsce. Woda – źródło życia*, Ekologia.pl, <https://www.ekologia.pl/srodowisko/specjalne/woda-zrodlo-zycia-rzecz-o-dostarczaniu-wody-do-miast-w-europie-i-polsce-w-dawnych-wiekach,15262,1.html> [Accessed: 20.02.2020].

Żylicz T., 2014, *Cena przyrody*, Wydawnictwo Ekonomia i Środowisko, Białystok.

Index of acronyms

ABC Waters	"Active, Beautiful, Clean Waters" programme of PUB in Singapore
BGI	blue-green infrastructure
BGN	blue-green network
CaBA	catchment based approach (Defra, UK)
CI	circular economy
EEA	European Environment Agency
FRM	flood risk map
FHM	flood hazard map
GI	green infrastructure
IPCC	Intergovernmental Panel on Climate Change
ISUDS	Integrated Sustainable Urban Development Strategies
IUWM	Integrated Urban Water Management
IWA	International Water Association
IWRM	Integrated Water Resources Management
LES	London Environment Strategy
LID	low-impact development
LRAP	London Rivers Action Plan
LWS	London Water Strategy
LSMP	local spatial management plan (in Poland)
NBS	Nature Based Solutions
PUB	Public Utilities Board in Singapore
RGB	Red–Green–Blue – structure of build-up (R), natural (G) and water (B) environments
SuDS	Sustainable (Urban) Drainage System
TRIP	pro-ecological storm water management system based on natural processes: transport (T), retention (R), infiltration (I) and purification (P) of water
URA	Urban Redevelopment Authority
WFD	Water Framework Directive (2000)
WSUD	Water-Sensitive Urban Design
WWC	Water-Wise Cities

www.ingramcontent.com/pod-product-compliance
Ingram Content Group UK Ltd.
Pitfield, Milton Keynes, MK11 3LW, UK
UKHW022131220326
11407UKWH00004B/27